C709

London 2007

D1330254

for the ...
built environment

B M Sadgrove MA CEng MICE

E Danson FinstCES FRICS (Revising author)

CIRIA *sharing knowledge* ■ *building best practice*

Classic House, 174–180 Old Street, London EC1V 9BP
TELEPHONE +44 (0)20 7549 3300 FAX +44 (0)20 7253 0523
EMAIL enquiries@ciria.org   WEBSITE www.ciria.org

Sadgrove, B M

Danson, E (Revision and update)

Edwin Danson is Senior Partner of Swan Consultants Ltd, an independent firm providing business development services, editorial and media support services as well as professional survey expertise and training to the international geospatial engineering community. Email: swanconsult@aol.com.

Setting-out procedures for the modern built environment CIRIA C709, 2007

First published 1988. Reprinted 1988, 1989, 1990, 1993, 1996
Second edition 1997, Reprinted 2003, 2004, 2005.

Revised and updated as *Setting-out procedures for the modern built environment* 2007

ISBN 0-86017-709-2   978-0-86017-709-8

Classification	
Availability:	Unrestricted
Content:	Advice/guidance
Status:	Committee guided
User:	Site engineers

Acknowledgements

The revision and update of *Setting-out procedures* (C707) was carried out by Edwin Danson, resulting in this new edition: *Setting-out procedures for the modern built environment* .

In preparing this revised and updated edition, the contributors included:

Steven Fox	*Edmund Nuttall Ltd*
Dr Gethin Roberts	*Department of Geomatics, Nottingham University*
Dr Paul Cruddace	*Geodetic Adviser, Ordnance Survey*
Edwin Danson	*Independent Consultant and editor*
Michael Sutton	*Environment Agency*
York Survey Supply Centre	*for the table on steel tape accuracy*
Mark Greaves,	*Geodetic Analyst, Ordnance Survey.*

CIRIA'S Research Manager for the project was Nick Bean.

Preface

Setting-out procedures for the modern built environment is an entirely revised and updated publication which includes modern instrumentation and techniques and describes recommendations for setting-out of building and civil engineering works. These are based on the practical experience of practising engineers and are applicable to most small to medium size construction contracts. On contracts involving high-rise structures or complex underground projects, state-of-the-art engineering design and concepts, unusual materials or robotic plant, the site engineer is encouraged to seek the advice and guidance of qualified and competent professionals.

Adoption of these procedures should reduce the incidence of errors, and the costs of putting them right. It is hoped that acceptance and use of these procedures will result in better communication and understanding between architects, consulting engineers, resident engineers and clerks of works, contractors' staff and foremen.

As in earlier editions, the epithet 'site engineer' is used throughout, but it is intended to apply to all site staff who are concerned with the process of construction setting-out. The old appellation 'chainman' has been updated to 'assistant'. It is assumed that these site staff will have a sound basic knowledge of surveying and are familiar with the more common instrumentation employed on construction sites and that the site engineer responsible for these activities is properly qualified and competent in setting-out works.

The site engineer is advised to report certain actions, errors and omissions. To whom the engineer should report is not generally stated because the reporting procedure will vary from site to site. Where the term 'supervisory authority' is used, this means the architect, engineer or supervising officer or a nominated representative of one of these.

Specific recommendations for the desired accuracy of setting-out have been avoided – any references to figures or dimensions are recommendations only or are used for clarity. Site engineers should set-out as accurately to the tolerances appropriate to the work, or specified for the construction, and choose instrumentation and techniques appropriate to meeting the criteria. If the instrumentation and resources provided to the site engineer are inadequate to meet the specification, then the site engineer must make this clear to the supervising authority in writing together with recommendations.

Visual and commonsense checks should always be applied. Although design details are normally not the contractor's responsibility, the site engineer is advised to check a number of such details, as best practice dictates, to smooth the running of the contract, minimise any consequential delay and encourage good relations.

The first edition of this publication superseded the CIRIA Manual of setting-out procedures and was based on the same material re-written with additions and rearranged in a format convenient for reference in the field. The second edition was updated in parallel with industry practice. This new version has been extensively revised and updated, particularly with the use of electronic instruments and computers.

Contents

Glossary

Assistant	Assistant to setting-out (site) engineer (in the past aka chainman)
Azimuth	Angle measured from True North
Bearing	Angle measured from Grid North
Benchmark	A height reference point or mark of known level
Boning (-in)	Sighting over profiles at travellers and establishing formation depths etc
Boning rod	Upright and rail forming a tee, 1 to 1.5 m high. Used when boning between two other boning rods of equal height
Borrow pit	Pit supplying soil for construction of a road or other foundation
Chainage	Distance along a line from an origin or datum point (formerly measured in chains, hence chainage)
Chord points	Points on a curve defining a chord
Collimation, line of	Optical axis of a telescope: also the line of sight through an instrument when set horizontal
Crossfall	Gradient on cross-section of road to shed water to one or both sides
Datum level	Horizontal plane of assumed (or actual) level from which other levels are determined
Datum	Horizontal, vertical or 3-dimensional coordinate reference definition
Deflection angle	Angle between the tangent and a chord at a point on a curve
Falsework	Temporary structure needed to construct permanent works
Forkhead	Scaffolding component supporting falsework beam
Formation level	Excavation level (height) on which permanent works constructed
Formwork	Forms used to constrain fluid concrete to a desired shape
Galileo	Developing European GNSS
GLONASS	Global'naya Navigatsionnaya Sputnikovaya Sistema (Russian GNSS)
GNSS	Global Navigation Satellite Systems (the generic term for satellite navigation systems, includes GPS, GLONASS and Galileo)
GPS	Global Positioning System (US GNSS)
Ground distance	True (actual) horizontal distance over the ground; the physical measurement of this distance (*see* also Projection distance)
HSE	Health, Safety and the Environment – a priority for all site engineers
Intersection point	Intersection of tangents from tangent points
Invert	Lowest internal level, at a given cross-section, of a pipe, channel or tunnel
Kicker	Concrete 'step' used to locate vertical forms
MOSS	Proprietary software for modelling three-dimensional surfaces
Partial coordinates	Coordinates of a point relative to another on the same grid. They are the algebraic differences between the eastings and northings
PPE	Personal Protective Equipment
ppm	Parts per million

Profile	A site marker delineating the shape/level etc of a construction element at that location
Projection distance	Distance *calculated* between coordinated points on a geodetic, eg National Grid projection (*see* also Ground distance)
Shoulder	Unpaved width at edge of road section
Sight rail	Horizontal or sloping rail of profile
Soffit	Lower surface of slab, bridge or similar structure
Summit	A high point on a road surface
Tangent point	Point defining start or finish of a curve
Temporary works	Temporary construction needed to construct permanent works
Total coordinates	Coordinates of a point referenced by eastings and northings relative to the origin of a grid
Total Station	A combined distance and angle (electronic) measuring survey instrument
Trammel Traveller	Rod used in describing arcs about a fixed centre for curved walls etc
Trench sheet	Steel sheet used in supporting trench sides
Tribrach	The three-armed support frame on which an instrument is mounted
Valley	A low point on a road surface
Whole Circle Bearing	Angle measured clockwise from *grid* North

Health and safety aspects

Health and safety of all those involved in construction work is a key consideration in the planning and management of site activities, including setting-out. Regulations which affect the planning and management of site work are constantly under review by government and other regulatory bodies, and it is imperative that the setting-out engineer is familiar with those regulations in force and relevant to his/her work.

Statistics indicate that those working on construction sites are most at risk during the first two weeks of their time on any new or unfamiliar site. Many of the hazards found on construction sites are explained in CIRIA's site guide SP130 *Site safety: a handbook for young construction professionals (2nd edition)*.

In an effort to improve safety, all personnel working on UK construction sites are now required to undertake formal site safety briefings and carry the Construction Skills Certification Scheme (CSCS) card, details of which can be found at: <www.cscs.uk.com>. The card lists the holder's qualifications and are valid for either three or five years. It also shows they have health and safety awareness as all cardholders have to pass the appropriate CITB-ConstructionSkills Health and Safety Test. Not having a CSCS card may affect your ability to work on certain sites.

Some typical hazards include:

- falls from heights or into excavations – edges should be protected with parapets or guard rails; use safety harness where appropriate
- falling objects – wear a safety helmet at all times; consider brimless helmet to avoid knocking instrument when taking readings
- rough ground and nails projecting from discarded timbers – wear approved safety footwear
- unsupported vertical excavations – do not enter trenches until supports have been installed
- confined spaces – ventilate to avoid poisonous gases or wear breathing apparatus
- working over or near water – wear approved lifejacket
- moving plant, lorries – wear high visibility jacket
- ladders damaged or not correctly fixed at top – see CIRIA publication SP121 *Temporary access to the workface: a handbook for young professionals*
- shot-fired nails – operator must be checked as competent – use protective glasses
- compressed air working in tunnels – conform to CIRIA Report 44, *see* Bibliography
- lasers – avoid looking into beams and take measures to prevent others from doing so.

In addition, general precautions can help prevent accidents:

- be aware of any existing site-specific Health and Safety Statement
- be aware of any existing site-specific Risk Assessments
- if one doesn't exist, establish an emergency plan – any vital contacts, nearest hospital A&E, etc
- ensure a properly equipped first aid kit is easily accessed
- ensure telephone contact is available, landline or mobile
- never cut corners on safety
- take immediate action to correct and report unsafe practices
- be aware of your surroundings and those of the assistant
- bring any health and safety considerations to the attention of anyone you bring to the site, contractors, consultants, etc
- be aware of actions that could affect other site workers, general public, etc
- keep aware of current and potentially changing weather conditions
- wear approved and suitable personal protective equipment (PPE) including UV protection
- do not run
- understand any COSHH assessments relating to chemicals you use – paint, sprays, etc

- never work alone:
  - in confined spaces
  - over or near water
  - on live electrical equipment
  - in derelict or dangerous buildings
- if possible avoid working alone:
  - on live roads
  - on roofs
  - in empty buildings
  - near demolition work
- whenever working alone:
  - use or establish a lone worker procedure
  - ensure a responsible contact is aware of your location, contact details and duration of stay
- never cut corners on safety
- take immediate action to correct or report unsafe practices.

Environmental aspects

The construction and demolition industry has consistently had a poor record regarding pollution of land and watercourses and the consequent damage to wildlife and natural habitats. The site engineer has a role to play in improving the environment by leadership and example.

Identify if the site owner/operator etc has an environmental and/or sustainability statement. Actively consider requiring or writing one; the paragraphs in this section would provide the essential content, however site-specific considerations also need to be taken into consideration.

By your actions, positively avoid causing unnecessary damage or harm to the environment and adopt an awareness of environmental matters to minimise the effects of your activities. Where possible, adopt best practice regarding all UK, EU and international legislative and regulatory requirements and agreements.

Where practicable, review activities and operations to identify environmental aspects and prioritise action to address them. Take a lead in raising awareness of environmental matters among employees, contractors, clients, suppliers, visitors, etc.

Where you have the influence, you should encourage minimising energy and resource consumption by promoting effective and efficient measures consistent with best practice. By influencing suppliers and contractors you should ensure that services and goods procured support national environmental policy.

Where possible influence and minimise the use of toxic materials and waste generated to prevent pollution. Dispose or recycle any waste in a responsible and appropriate manner. Ensure good management practices by reviewing them regularly, to verify their effectiveness in achieving environmental gains.

Surveying and setting-out instruments

Type	Remarks
Theodolite	Optical/manual (analogue) instrument for the measurement of angles only
Total Station	Most common electronic instrument used on site for the measurement of both angles and distances. Measurement information displayed digitally can be stored in a data logger.
	The Total Station includes a built-in computer that provides the site engineer with a number of features that typically include: <ul><li>input of ambient metrological conditions</li><li>observed positions as Cartesian coordinates or bearing and distances</li><li>distances corrected for slope</li><li>setting-out mode either using either coordinates, offsets or bearing and distances.</li></ul>
	Reflectorless Total Stations do not need a prism reflector set at the target. Robotic Total Stations do not need an operator at the instrument.
Gyroscopic theodolite	For measuring and setting-out angles relative to True North – especially useful when working underground.
GPS	Generic term for a range of positioning and setting-out solutions using the US DoD's NavStar Global Positioning System (GPS) satellites.
	For the site engineer, the use of GPS for control is the most relevant application and is discussed in the section Global Positioning System (GPS). Many GPS systems now also incorporate the GLONASS satellite navigation system.
	It should also be noted that GPS aided/robotic systems are becoming increasingly common for the automatic control of earth moving machinery. This aspect is outside the scope of this book.
Optical level	Optical/manual instrument only. Suitable for most site applications.
Automatic level	Similar to the optical level in principle but susceptible to vibration.
Digital level	Used with a 'bar-coded' staff for precise levelling.
Precise level	Also known as a geodetic level. Only for very high accuracy control and requires a competent specialist operator.
Optical plumb	Optical/manual instrument only. Average of four readings should be taken with instrument turned horizontally 90° between readings.
	For automatic version, normally two readings separated by 180° are sufficient.
Lasers	Alignment – used to define a line/direction Rotating (horizontal) – defines a horizontal plane Rotating (general) – defines any set plane.
	Pipe – defines line and grade.
Optical square	For setting-out right angles over short distances only.

Table 1   *Example of common surveying and setting out instrumentation*

Surveying instruments

It is assumed the site engineer is familiar with the use of the more common instruments used in setting-out works.

A sample of the more frequently encountered instruments is listed in *Table 1 Surveying instruments,* although the list changes frequently as new and innovative instruments come onto the market.

Although setting-out is greatly facilitated by modern instrumentation, their potential high accuracy should not be taken for granted and they must always be checked before use. Furthermore, accurate setting-out of works can still be achieved with less sophisticated instruments although the task may take longer. The choice of which instruments to use depends upon many factors including:

- size of the site
- complexity of the work
- precision/accuracy demanded
- economics: the time a task requires may be a dominating factor.

The manufacturer's instructions must be studied and followed. Inexperienced site engineers should take every opportunity to work with more experienced engineers.

See also Appendix B: Care and checking of equipment.

Setting-out techniques

Horizontal

Total stations and GPS receivers are both commonly used for setting-out, and often combined. The coordinates of points for setting-out can be uploaded directly from the design software into the equipment and then software tools are displayed, which allow the surveyor to locate the points to be set out. Total Station/GPS equipment has become very sophisticated, and can provide the engineer with useful information, but there is still need to have a full understanding of the keyboard operation and the correction factors and errors associated with the instruments.

Theodolite and tape is still a common method for setting-out. Care must be taken to apply corrections to all taped distances, however short. For accuracy, it is essential to check any angle or distance that has been set out. This can be done by setting up the instrument at key positions and observing all related points and comparing these angles with angles calculated from construction drawing dimensions. This will locate any errors and confirm that the specification has been met.

A Total Station can be used to set out many points from one station, working on a coordinate system or by bearings and distances. This method speeds up the setting-out process while maintaining a high order of accuracy. Total Station technology is continually advancing and these instruments provide the site engineer with useful information. Manufacturer's instructions must be studied carefully to understand the keyboard operation, the capabilities (and limitations) of the instrument, the application of correction factors and to reduce the potential for inadvertent error.

The cost of renting Total Stations has decreased greatly since their introduction, as has the cost of the instruments themselves.

The choice of traditional or modern systems comes down to the following:

Theodolite and tape

- low cost equipment
- simple to use
- requires two people – operator and assistant
- slow operations compared to Total Station
- excellent for small, simple sites.

Total Station

- reasonably inexpensive to hire
- requires familiarisation with the functions and operation of the instrument
- efficient setting-out of multiple points
- excellent for larger scale/more complex sites
- requires a battery recharging supply.

Versions available include:
- reflectorless (doing away with need for a pole mounted prism)
- robotic (one-man operation).

Equipment suppliers should be consulted for a more detailed analysis of the benefits of the two methods. Whichever type of instrument is used, daily checks should be made to ensure that the instruments are working properly, and all setting-out points must be cross-checked and the results recorded to show that the specifications have been met.

Vertical

Optical tilting or automatic levels are adequate for most setting-out purposes and site control where sight lines can be kept reasonably short. For more advanced applications, a precision level is required for example when:

- establishing primary benchmarks for an extensive site or long overland site, eg railway or road
- monitoring movement of structures
- track laying to high tolerances.

The digital version of the precise level has further advantages:

- speed of setting up
- immediate calculation of reduced levels
- storage of information in data logger
- automatic compensation.

Setting-out control

Setting-out control is normally divided into three Orders:

1 **Primary or First Order**: the network of control points coordinated in three dimensions to provide an overall fixed reference framework for the works

2 **Secondary or Second Order**: local control points for construction or baselines established for setting-out the infrastructure

3 **Tertiary or Third Order**: points set out at the actual positions for construction or at suitable offset positions.

Each level of control is verified and agreed before proceeding to the next. Points at any level are established and checked from points of a higher order.

Refer also to BS5964/ISO 4463 – 1979, – Setting out and measurement of buildings.

Precision naturally decreases with each order therefore, to maintain the accuracy of the actual setting-out point demanded in the specification, the precision of the preceding orders must be higher. For example, if the required accuracy of a (tertiary) point within the framework of the site reference system were 2 cm, then the accuracy of the secondary control station that it was set out from would have to be better than 1 cm.

Accuracy

The final accuracy required for the setting out of works will be determined by the specification. Greater accuracy and care may be necessary in parts of the work to achieve this, eg for a fit between one part and the next. The need for economy in construction can also be a factor in deciding the extent of the setting-out, but the site engineer is cautioned that a parsimonious approach will inevitably lead to poor work, unnecessary time delays and increased costs due to remediation.

Linear measurements

Conventional steel tapes

The precision achievable with slope correction and the appropriate tension applied (if known) is circa. 0.01–0.05% of distance measured, eg 1 to 5 mm over 10 m.

York Survey Supply Centre, provides the following useful table on its website <www.YorkSurvey.co.uk>

Accuracy & Tolerances
The tape blade length accuracy is given by the formula:

$$ECI \quad \pm [0.1 \text{ mm} + (L \times 0.1 \text{ mm})])$$

$$ECII \quad \pm [0.3 \text{ mm} + (L \times 0.2 \text{ mm})]) \text{ where } L = \text{Length rounded up to the next whole metre above.}$$

$$ECIII \quad \pm [0.6 \text{ mm} + (L \times 0.4 \text{ mm})])$$

The tolerance applies at the temperature and tension printed on the blade. If no temperature or tension is specified then the tolerance applies at a temperature of 20°C and zero tension.

Tolerances for typical blade lengths are:

Length	ECI ±mm	ECII ±mm	ECIII ±mm
1 m	0.2	0.5	1
2 m	0.3	0.7	1.4
3 m	0.4	0.9	1.8
3.5 m	0.5	–	2
5 m	0.6	1.3	2.6
6 m	0.7	1.5	3
8 m	0.9	1.9	3.8
10 m	1.1	2.3	4.6
15 m	1.6	3.3	6.6
20 m	2.1	4.3	8.6
25 m	2.6	5.3	10.6
30 m	3.1	6.3	12.6
50 m	5.1	10.3	20.6
100 m	10.1	20.3	40.6

Specialist tapes

The use of a certified and calibrated steel bands with all corrections applied (tension, temperature, slope {and catenary}) can provide accuracy of a much higher order, circa. 0.002%, but requires expert operation.

Total Station

Distance accuracy depends on the quality of the instrument but typical values are 1 mm ± 5 ppm eg 1 mm for the shortest distance, ranging up to 1.5–2 mm over 100 m.

Absolute and relative accuracy

Absolute accuracy refers to the precision of individual points within a control framework. For example, a point located by a differentially corrected GNSS receiver within, say, the Ordnance Survey National Grid, may be ±2 cm. This does not mean that, relatively, all such points are within ±2 cm of each other.

Relative accuracy refers to the precision of measurements of one or more points to each other. In the above example, the relative distance between the two GNSS points, while precise in absolute terms, could range from 0 cm to 4 cm, ie ±2 cm + ±2 cm. The distinction is particularly important when setting out using GNSS systems but equally applies to other instruments.

Levels

Optical level: (builder's type) ± 5 mm over 30–50 m
                   (engineer's type) ± 2mm over 30–50 m

Digital/precise levels can be read to 0.1 mm with precision in the order of twice this.

A run of levels should open and close on different TBMs to eliminate the common error of using an incorrect starting level. For tertiary points, the level run should close better than $1.5 \times \sqrt{2n}$ mm where n is the number of set-ups. Example: a level run of nine set-ups = $1.5 \times 9 = 1.5 \times 3 = 4.5$ mm maximum misclosure.

Horizontal angles

20" theodolite: ± 20" or 10 mm in 100 m
5" theodolite: ± 5" or 2.5 mm in 100 m

These estimates assume the average of observations on each face with readings in opposite quadrants of the circle. If the difference is twice the nominal accuracy of the instrument, the observations should be repeated. If the error remains, then the instrument needs checking.

Total Station (typical): <3" or <2 mm/100 m

Typically, Total Stations are self-calibrating (see manufacturer's instructions) and do not require readings on opposite faces. However, the prudent engineer will check this first.

Vertical alignment

Plumb-bob in still conditions:

- Freely suspended: ±5 mm in 5 m (maximum recommended distance)
- Oil damped: ±5 mm in 10 m (maximum recommended distance)

Optical plummet: (properly adjusted and operated) ±5 mm in 100 m
Laser plummet: (properly adjusted and calibrated) ±4–8 mm in 100 m.

Note: The nominal accuracy of an instrument, or the precision with which it can be read, is not the same as the accuracy of the actual measurement which is affected by corrections factors and others including, for example, rounding errors or operators personal judgement.

Checks: physical and calculation

The site engineer must acquire or develop methods of checking and cross-checking all setting-out operations until they become second nature. For particularly critical setting-out, seek an independent check by someone else. Where this is not practicable self-checking is essential.

When principal lines and levels have been set out, the contractor should advise the supervising authority. The relevant person may arrange an independent check of these lines and levels, but the site engineer should not count on this. The responsibility for the setting-out will remain with the contractor.

Physical checks

Make physical checks. Where possible check:

- visually that lines and levels tie in with existing features
- distances to nearest metre by rough taping or pacing
- levels by 'eyeing-in' on known levels
- that supposed right angles look to be correct
- that falls are in the right direction
- verticality approximately with spirit level.

Calculation checks

Calculations must always be checked. Most site engineers use calculators or PCs and tend to rely on them implicitly. It is, however, all too easy to input the wrong figures or press the wrong function button or apply the wrong sign. Therefore a check or independent calculation is imperative. Where possible, data for check calculations should be input in a different order to minimise the risk of mis-keying an incorrect value twice. For example, if adding a column of figures, input from the top for the first total and then from the bottom as a check.

Conversions from degrees, minutes and seconds to decimal degrees are helpful for various computations but must be carefully executed. The formula is degrees + minutes/60 + seconds/3600.

Date .... Levels taken for ....

From .... To ....

Back-sight	Inter-sight	Fore-sight	Rise	Fall	Reduced level	Distance	Remarks

LEVEL BOOK HEADINGS - 'RISE-AND-FALL'

Back-sight	Inter-sight	Fore-sight	Height of collimation	Reduced level	Distance	Remarks

LEVEL BOOK HEADINGS - 'HEIGHT OF COLLIMATION'

BLOGGS CONTRACTORS LTD   No. 00001
Old Wharf
Newtown

SITE INFORMATION SHEET

Copies to: .. General Foreman
Date .... Ganger
Contract .... Site Office (Engs)
Compiled by .... Site Office (Q.S.)
                                    .. Resident Engineer

SITE INFORMATION SHEET

Recording and informing

Keeping records

Good records are essential for:

- accurate construction
- dissemination of setting-out information
- accurate measurement of completed work
- resolving disputes with supervising authority.

Records to be maintained during the job and retained until all contractual obligations are fulfilled include:

- all drawings issued by the supervising authority
- site instructions issued by the supervising authority
- coordinates of main setting-out points
- locations and levels of relevant benchmarks and original ground levels
- records of located/relocated underground and over-ground utility services
- digital photographs/sketches of obstructions not shown on original drawings
- classes of subsoil encountered and obstacles to excavation
- printouts and calculations of the setting-out
- all field books and original observations
- computer files
- copies of Site Information Sheets received and issued
- CAD/hardcopy scale drawings of as-built works.

It is prudent to duplicate key records, especially those relating to measurement, and to store these records and any computer back-up files separately off site.

Field documents

Site engineers should be supplied with:

- level books ruled for 'rise and fall' or 'height of collimation'
- observation/record books for use with theodolites, Total Stations or other site instrumentation
- pads of Site Information Sheets, preferably sequentially numbered
- pads of sewer information sheets.

Site Information Sheets

The site engineer should use Site Information Sheets to confirm and supplement oral information/instructions given to foremen and gangers etc. Use sketches to clarify the information as appropriate. Typical details to be included:

- essential dimensions
- offset distances from line pegs
- levels of offset level pegs
- levels of profiles and length of travellers
- spoil heaps, borrow pits and haul roads.

Quality assurance

Construction sites have quality systems in place that may be independently audited in accordance with the Quality Assurance procedures or by the supervising authority or end customer. These procedures will establish the types of information and details that need to be recorded from any setting-out. They may be company standards or tailored to the needs of a specific site.

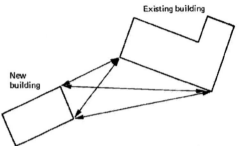

Check that location of the works is adequately shown on the drawings

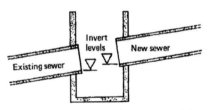

Check that existing and new levels are compatible

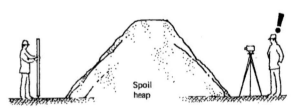

Temporary features can obstruct setting-out!

Initial actions

Before starting to set out the works, the following actions need to be taken:

1. Review and become familiar with the drawings and check that:
   - the drawings are the latest issues
   - all essential dimensions are given (do not scale) and that intermediate dimensions agree with overall dimensions
   - the location of the works, in relation to permanent or temporary reference points, is adequately shown on the drawings
   - the level heights of the works, in relation to permanent features or temporary benchmarks, is shown
   - critical dimensions between related components are clearly indicated.

2. Walk over the site checking that:
   - boundaries are well-defined and are as shown on the drawings
   - all permanent and temporary reference points and benchmarks are as shown on the drawings
   - all visible permanent features are correctly indicated on the drawings
   - any permanent or temporary features do not interfere with the setting-out or construction of the works
   - there is no evidence of hidden features that might affect the setting-out or construction of the works – (also check the contract documents for any warnings or cautions).

3. Report any discrepancies found from the above checks in writing and keep a copy. It may take some time to rectify discrepancies.

4. Confirm any oral instructions in writing.

5. Set up a system of recording and communicating information.

6. Set up the CAD/computer-aided construction system. Check the files for function.

7. Begin to train the setting-out team (if necessary).

8. Set up and/or prove setting-out stations.

9. Set up and/or prove site benchmarks.

10. Record and agree existing site levels and features.

11. Check that the works will tie in with any existing works.

12. Plan the sequence of setting-out and how dimensions will be controlled.

13. Make sure you are able and equipped to perform the work, for example are the instruments adequate, in good working order and fully adjusted? Is your assistant properly trained? Are you properly trained? Have you sufficient resources?

14. Do your thinking in the office before getting to the site.

Note: Report immediately any errors in setting-out as soon as they are discovered. Early action to correct errors will minimise costs. Never conceal your errors or your doubts.

Move top of
staff in
direction shown

Reading taken
and recorded
(thumb OK)

Drive peg in
at least 25 mm

Tap peg in
about 3–5 mm
(pinch)

Move pointer
in direction
indicated

Pointer on line:
mark or tap in
nail (thumb OK)

Instructing the site assistant

A good assistant is an essential member of the setting-out team, contributing considerably towards the speed and accuracy of the setting-out work. The site engineer frequently has only the services of an inexperienced person and it is good investment to devote time to training the person before attempting any critical setting-out.

On small sites, the amount of setting-out may not justify a full-time assistant and the site engineer has to 'borrow' someone. In this case, try to use the same person each time.

Check on the experience of the assistant. Where necessary, explain the basic principles of setting-out, stressing the importance of accuracy and the role of the assistant in achieving accuracy and speed of setting-out.

Topics that may need to be covered include:

- safety aspects
- what the various instruments are for
- constructing and protecting setting-out stations
- measuring with tapes
- constructing and checking temporary benchmarks (TBMs)
- use of level staff and setting a peg to level
- setting up and using profiles
- using plumb-bob or optical plumbing instrument
- care of equipment
- maintaining stocks of pegs, nails, paint and other stores.

Signals and signs

The page opposite illustrates some of the common signals used to communicate information between site engineer and assistant. Mobile phones or low wattage radios can also be used subject to site conditions and restrictions.

Responsibility for the assistant

The site engineer has a responsibility under health and safety legislation to take reasonable care for the health and safety of those affected by his/her actions, including any assistants. Ensure that the assistant is aware of the potential hazards of the site and is using the correct personal protective equipment and clothing.

Assistants have been seriously injured and even killed by passing plant and traffic. Before signalling the assistant to move, check that it is safe to do so.

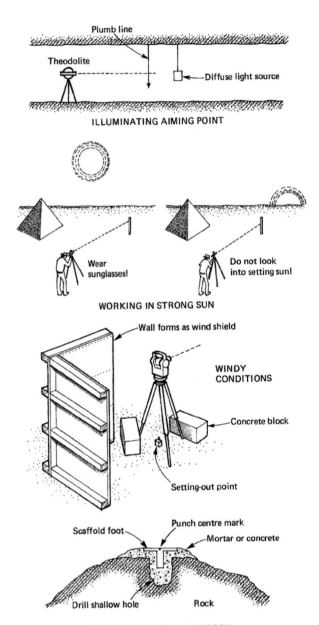

Plumb line

Theodolite

Diffuse light source

ILLUMINATING AIMING POINT

Wear sunglasses!

Do not look into setting sun!

WORKING IN STRONG SUN

Wall forms as wind shield

WINDY CONDITIONS

Concrete block

Setting-out point

Scaffold foot

Punch centre mark

Mortar or concrete

Drill shallow hole   Rock

SETTING-OUT STATION ON ROCK

Difficult ambient conditions

Poor light

- use surveying instrument with integral lighting system to illuminate scales and cross-hairs. In the absence of a lighting kit, a pen light suffused by tissue paper will suffice. A little light aimed towards the object lens (not directly) will aid in illuminating cross hairs
- use diffuse light source to illuminate staff or aiming point (for example, place source behind plumb line, etc)
- GPS systems are normally equipped with illuminated displays and can be used in poor lighting conditions.

Strong sun

- shade instrument to avoid bubble disturbance – this is less of a problem with automatic self-collimating instruments
- wear good quality sunglasses in strong sunlight or use a light filter over the eyepiece
- to avoid heat shimmer, set out when sun is low but never look directly at the sun through the telescope.

Cold weather

- wear warm wind-proof and waterproof clothing
- wear mittens for ease of adjusting instruments.

Windy conditions

- ensure feet of tripod are suitably weighted down
- in extreme conditions, shield instrument.

Salt spray or dust

- meticulously clean instruments each day
- clean lenses with recommended tissue or brush only
- check for bearing wear on instruments frequently.

Noisy conditions

- carry out all preliminary calculations in site office to avoid being confused by noise
- brief assistant fully before going into a noisy area
- subject to the need to hear alarms, wear ear defenders.

Vibration

- position instruments as far from vibration sources as is practicable
- set-out before operations start or during meal breaks
- ensure readings are consistent before accepting them.

Soft ground

- for setting-out stations or benchmarks use existing features/works wherever possible, construct mass concrete block or install pile
- use offsets liberally – be resigned to resetting pegs frequently
- check pegs from offsets, do not assume they are correct or remain undisturbed
- avoid moving around tripod more than necessary.

Rock/hard ground

- use rock drill to make hole for level peg or steel pin
- for setting-out stations, use scaffold foot as shown in diagram.

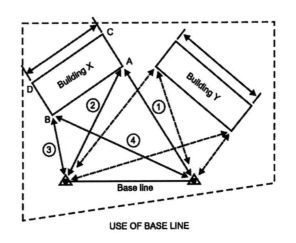

USE OF BASE LINE

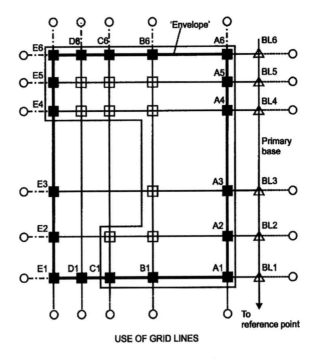

USE OF GRID LINES

Base and grid lines

Although grid coordinates are increasingly common for setting-out buildings, baselines and (setting-out) grid lines are still commonly used. If this is the case, the contract drawings should indicate the baseline or grid lines to be used.

A baseline is suitable for small sites where referencing back to the baseline for all the works is possible. Grid lines are preferable for larger sites where the baseline may be obscured by the works as they progress or where referencing would be cumbersome. Whichever is used, the site engineer should prove any setting-out stations already provided or construct the stations and agree these with the supervising authority.

Use of baseline

A baseline comprises two setting-out stations a given distance apart. In the example, point A of building X is set out by taping dimensions 1 and 2 from the baseline and point B is set out by taping dimensions 3 and 4.

The dimension AB is then checked against that required. Provided there is no anomaly, the remainder of the building can be set out from AB, which effectively becomes the baseline for the building.

Corner profiles can then be set out for the building.

Use of grid lines

Usually, a primary baseline is provided from which all other positions can be set out. It is normal to establish a rectangular grid that includes all sides of the structure (*see* example). The corner positions and the centres of columns and their bases are set out from the baseline.

Each corner is checked by theodolite or Total Station, comparing observed and calculated angles. Any adjustment to the points to meet specification is done at this stage. The grid lines should then be projected beyond the construction area and referenced to well-constructed/permanent control stations. Alternatively, subject to the accuracy required, a GNSS receiver can be used to define the base line and subsequent grid lines (NB note the comments on absolute v relative position).

Example of checking procedure (see opposite)

- set up instrument on BL1 and zero to BL6
- observe A1, B1, C1, D1, E1, E2, E3, E4, E5, E6, D6, C6, B6, A6, A5, A4, A3, A2
- set up instrument on A1, E1, E6, A6, BL6 and repeat
- adjust points if necessary
- reference lines beyond construction area.

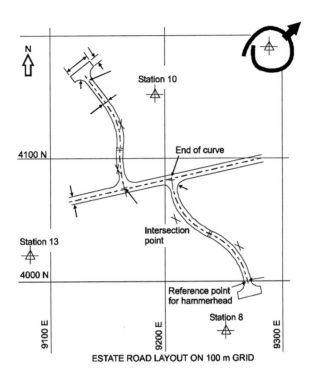

N

Station 10

4100 N

End of curve

Intersection
point

Station 13

4000 N

Reference point
for hammerhead

Station 8

9100 E

9200 E

9300 E

ESTATE ROAD LAYOUT ON 100 m GRID

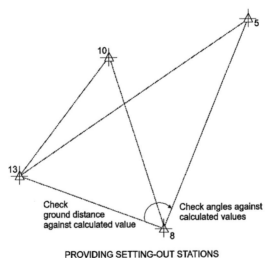

5

10

13

Check
ground distance
against calculated value

Check angles against
calculated values

8

PROVIDING SETTING-OUT STATIONS

Grid coordinates

The ease of providing setting-out coordinates using a Total Station or GPS receiver, has meant that setting-out by coordinates has become relatively straightforward. The grid for the coordinates may be set up specifically for the site or may, in the case of large linear works, use the Ordnance Survey National Grid (or other national geodetic frameworks). The general principles will be the same but note that when the National Grid is used, distances must be corrected to allow for projection scale factor (*see* Appendix C: UK National Grid, benchmarks and ground distances).

Whatever grid is used, setting-out stations will be provided or will have to be established and constructed, but not necessarily lie on the main grid lines. The estate road layout (opposite) illustrates this.

Before starting setting-out, walk over the site and inspect the setting-out stations for signs of damage or displacement. Report any suspect stations in writing.

Grid coordinates by Total Stations

The process is greatly speeded up, and is likely to be more accurate, by using a Total Station where coordinates are displayed directly. Before defining new points, check equipment is functioning correctly by checking points of known coordinates. Prove the stations as follows:

- set up on a station of known coordinates and RO to another given station
- observe the other stations in turn and compare their observed coordinates with their given coordinates.

Grid coordinates by bearing and distance

If using a theodolite and tape, assuming that stations appear satisfactory, prove the stations as follows:

- calculate ground distances between stations from coordinates
- check ground distances on site (correct measured distance for slope, as required)
- assume that two stations the correct distance apart, define a baseline
- calculate angles between this baseline and the remaining stations
- check these angles.

The distances and relative bearings of points to be set out should next be calculated and then applied on the site, correcting distances for slope as necessary.

General

- report in writing any discrepancies outside specified limits
- agree discrepancies and actions with supervising authority
- do not rely on a single calculation. Repeat calculation, using different order of input if possible, or ask somebody else to check the calculation.

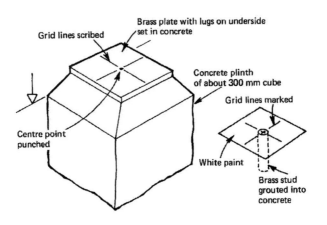

Grid lines scribed

Brass plate with lugs on underside set in concrete

Concrete plinth of about 300 mm cube

Grid lines marked

Centre point punched

White paint

Brass stud grouted into concrete

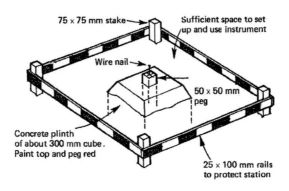

75 × 75 mm stake

Sufficient space to set up and use instrument

Wire nail

50 × 50 mm peg

Concrete plinth of about 300 mm cube. Paint top and peg red

25 × 100 mm rails to protect station

PRIMARY SETTING-OUT STATIONS

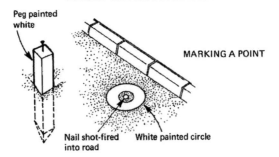

Peg painted white

MARKING A POINT

Nail shot-fired into road

White painted circle

Lines

Location of control points

Control stations provide a coordinated position (Easting, Northing) and normally include a referenced level (TBM).

Depending on the accuracy required, stations may be very sophisticated and accurately located or as simple as a nail shot-fired into a blacktop surface.

Careful planning is required for the location of control stations. Ideally, they should be located in areas where they are unlikely to be disturbed by the works and remain in place until the completion of the project.

Therefore:

- check all drawings to identify safe areas for control stations
- ensure control stations are sited clear of permanent and temporary works
- provide good access to allow for an instrument to be set up safely
- ensure the location provides clear lines of sight to other points
- allow sufficient room to move around the instrument
- secure the control station peg or marker well into ground, with a nail marking the exact spot
- protect the station with four stakes and attach rails (see diagram)
- set concrete plinth with station number inscribed.

For success, the prime positions chosen will need to be set out with taped measurements. Control stations for interior works require the same degree of care and should be marked out at an early stage, when access and line of sight is best and usually when the structural frame and floor slabs are in position.

Marking a line

When setting-out a new line:

- turn the required angle of the line on both faces of the instrument and make two temporary marks
- subject to marks being within acceptable tolerance, take the mean as being on new line
- make permanent mark. Delete temporary marks.

The new line may be defined by one or more points (in addition to the original baseline station) marked sufficiently to meet the needs of the work and permanence.

Indicating a line

Ranging rods are commonly used to temporarily indicate a line, eg for earthworks, where precision is not critical. On no account should timber battens be used for ranging purposes as this causes confusion.

Offset pegs

In many cases, station, points and lines set out from the setting-out stations will be lost or destroyed during the construction process. To allow original points and lines to be redefined offset pegs are necessary.

Redundant markers

Obsolete or disturbed setting-out stations, pegs or other markers should be removed to reduce risk of setting-out errors and to avoid endangering the public once a site is vacated. Recycle or dispose of materials in a responsible manner.

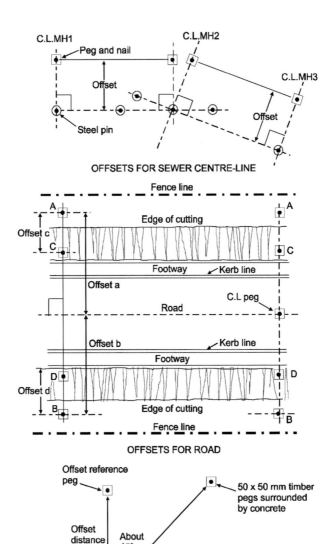

C.L.MH1
Peg and nail
C.L.MH2
C.L.MH3
Offset
Offset
Steel pin

OFFSETS FOR SEWER CENTRE-LINE

Fence line
A
A
Edge of cutting
Offset c
C
C
Footway
Kerb line
Offset a
C.L peg
Road
Offset b
Kerb line
Footway
D
D
Offset d
B
Edge of cutting
B
Fence line

OFFSETS FOR ROAD

Offset reference
peg
50 x 50 mm timber
pegs surrounded
by concrete
Offset
distance
About
45°
About 45°
At least three
reference pegs
Setting out point

OFFSET REFERENCES

Offset pegs

Need for offset pegs

Original setting-out lines (eg centre-lines) and setting-out points (eg the centre of a circular structure) are frequently destroyed, disturbed or obscured as the construction process moves forward. Installing offset pegs allows the original lines and points to be re-established.

Work out where offset pegs will be needed in good time and install them as soon as possible.

Offset distances

On a slope, use a short offset distance to minimise any error in correcting for the slope. In general, keep all offset distances short, consistent with the pegs being reasonably secure against disturbance.

All offset pegs should be located within the site boundary.

Offsets points for sewer centre-line

Typically, replacement sewers are set out using steel pins to mark the centres of manholes. Additional setting-out points on the line are added as required. Offset pegs are then positioned perpendicular to the centre-line and located in an area least liable to disturbance from the trenching (or other construction) process. The distance from the centre-line to the offset peg is measured; preferably, the offset distances should be kept the same. A simple method of offsetting a right angle is adequate, eg with an optical square. Details of the original setting-out points, offset pegs, and offset distances should be recorded on a sewer information sheet.

Offsets for road

The centre-line pegs are set out first. In the diagram, offset pegs A and B are positioned beyond the edge of the earthworks but within the fence lines. The offset distances *a* and *b* may not be equal if the earthworks are not symmetrical. Profiles for the earthworks can then be set out and/or re-established from these offset pegs.

When the earthworks are complete, offset pegs C and D can be positioned clear of the outside line of the footway using offsets *c* and *d* from A and B respectively. Offset pegs C and D are used to set out the kerb line.

Offset reference pegs

For a critical setting-out point, at least three offset pegs should be used in the event that one is disturbed or destroyed.

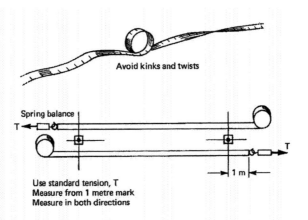

Avoid kinks and twists

Spring balance

T

1 m

Use standard tension, T
Measure from 1 metre mark
Measure in both directions

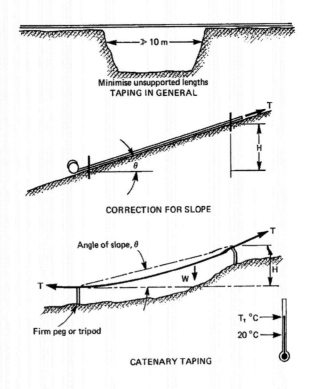

≥ 10 m

Minimise unsupported lengths
TAPING IN GENERAL

T

H

θ

CORRECTION FOR SLOPE

Angle of slope, θ

T

W

H

T

Firm peg or tripod

T_t °C

20 °C

CATENARY TAPING

Measurement with tapes

Taping remains one of the chief tools of the setting-out engineer. For efficient work, the right tape for the job is needed and, for accuracy, the proper procedures observed.

Choice of tapes

- for setting-out accurately over short to medium distances, use only steel tapes to BS 4035 standard
- use plastic or steel reinforced plastic tapes only for approximate measurements
- use invar bands for special precision, eg in tunnels.

Note: Set aside and label a BS 4035 standard steel tape a 'standard tape' to be used only for checking other (working) tapes.

Care of tapes

- clean the tape before rewinding into case
- clean and lightly oil steel tapes at end of working day
- check (working) tapes against standard tape each week and after repair.

General measuring with steel tape

- ensure steel tapes are not kinked or twisted
- where possible, measure over level ground, avoiding stones, tree roots etc
- measure close to the ground to minimise wind disturbance and unsuspended sections
- for precise measurements, apply temperature corrections and the standard tension marked on tape
- avoid unsupported spans exceeding more than 2 m
- check overall length against sum of intermediate lengths
- take mean of two measurements in opposite directions, corrected for slope, temperature, sag
- to avoid errors in critical measures, do not use the tape zero. Instead, measure from a point close to the 1 m mark, note the value, and adjust the end reading accordingly. Do this at least twice. Be sure to instruct the assistant in this method of 'slip chaining'
- align intermediate measuring stations carefully by instrument or eye.

Correction for slope

Distance along slope = L (m), difference in level = H (m)
Horizontal distance = $\sqrt{(L^2 - H^2)}$

Correction for temperature

The length of a steel tape will change with temperature, expanding with heat and contracting with cold. As tape *expands*, the *measurement* appears to *decrease* and the correction is *positive*. When *setting-o–ut*, the correction is *negative*.

As tape *contracts*, the *measurement* appears to *increase* and the correction is *negative*. When *setting-out*, the correction is *positive*.

Reading on tape = L(m); Temperature of tape = T_t °C; Calibration temperature of tape = 20°C

Correction to be applied = $11 \times 10^{-6} L(T_t° - 20°)$ (m), ie for 10°C difference in temperature the correction = 11 mm over 100 m.

Correction for sag (catenary taping)

Unless the site engineer is familiar with catenary taping, this method should always be performed by a qualified expert.

Measured length = L (m); Angle of slope between supports = $\theta°$; Weight/unit length of tape = W (kg/m); Tension applied to tape = T (kgf) (ideally standard as above)

Corrected length = $\dfrac{L - W^2 L^3 \cos^2 \theta}{24T^2}$

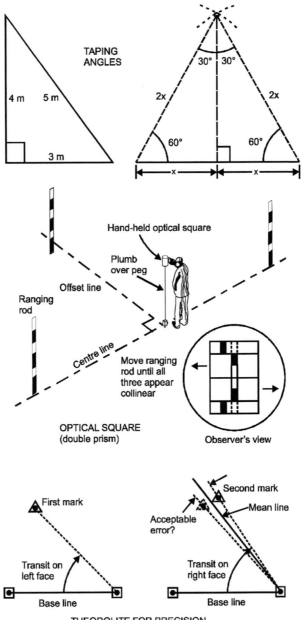

TAPING ANGLES

4 m 5 m

3 m

30° 30°

2x   2x

60°   60°

x   x

Hand-held optical square

Plumb over peg

Offset line

Ranging rod

Centre line

Move ranging rod until all three appear collinear

OPTICAL SQUARE
(double prism)

Observer's view

First mark

Transit on left face

Base line

Second mark

Mean line

Acceptable error?

Transit on right face

Base line

THEODOLITE FOR PRECISION

Horizontal angles

Simple techniques

Simple techniques are adequate for:

- setting-out offset pegs from centre-lines
- indicating approximate limits of excavation
- generally, where accuracy is not critical.

Such techniques include use of tapes, set squares and optical squares.

Using tapes

This technique can be used for angles of 30°, 60° and 90° (*see* illustration opposite) but does require:

- a reasonably level site
- generally two tapes and extra pegs.

Using a set square

A site-made set square can be handy for squaring up for small excavations or setting-out offset pegs fairly close to a centre-line.

Using an optical square

A hand-held optical square is a convenient instrument for setting-out right angles to offset pegs where an error of a degree or two has no significant effect on the offset distance. The maximum sighting distance should be limited to between 15 to 20 m. The method is shown in the illustration.

Using a theodolite or Total Station

For accurate setting-out with a theodolite or Total Station, remember to:

- transit with both face left and face right, using different zeros
- check, through assistant, that error is acceptable to allow mean to be taken
- ensure mean point is clearly marked.

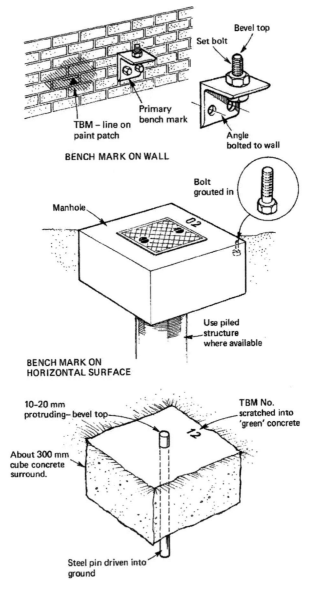

Bevel top
Set bolt
Primary bench mark
TBM – line on paint patch
Angle bolted to wall

BENCH MARK ON WALL

Bolt grouted in
Manhole
Use piled structure where available

BENCH MARK ON HORIZONTAL SURFACE

10–20 mm protruding– bevel top
TBM No. scratched into 'green' concrete
About 300 mm cube concrete surround.
Steel pin driven into ground

TEMPORARY BENCH MARK IN FIRM GROUND

Benchmarks

Each site should have a primary benchmark, which may be:

- a convenient nearby Ordnance Survey benchmark (but see note)
- referenced to an Ordnance Survey benchmark (but see note)
- referenced to a given level on existing works
- a height above Ordnance Survey Datum.

Note: Traditional Ordnance Survey benchmarks are no longer supported or maintained by the Ordnance Survey and may be unreliable. For establishing levels relative to Ordnance Datum, *see* Appendix C: UK National Grid, benchmarks and ground distances.

The supervising authority should specify which method applies. If the site levels are to relate to Ordnance Survey benchmarks but a particular benchmark has not been specified, select at least two local benchmarks and double level between them via at least two TBMs on the site. If height above Ordnance Datum Newlyn (or other Ordnance Survey height datum when working in some Scottish islands), the procedure outlined in Appendix C: UK National Grid, benchmarks and ground distances, should be followed.

Unless all levelling can conveniently be referred to the primary benchmark, it will be necessary to set up secondary benchmarks. Secondary benchmarks are commonly called temporary benchmarks (TBMs).

To ensure accurate primary and temporary benchmarks:

- establish primary benchmark from the agreed Ordnance Survey benchmark(s)/National Vertical Reference datum or from existing works and agree the level with the supervising authority in writing
- plan the locations of TBMs in good time, taking account of temporary and permanent works. All points of the works should, where possible, be within 40 m of a TBM and there should be two, independent, TBMs accessible for all potential instrument set up positions
- verify the levels of previously established TBMs by levelling from the primary benchmark
- establish TBMs not more than 80 m apart. Closing error to primary benchmark must not exceed $1.5 \times \sqrt{2n}$ mm (where *n* is the number of set-ups) or whatever is specified in the contract
- use existing permanent features for establishing benchmarks whenever possible (*see* illustrations)
- where no permanent feature is available, establish the benchmark in firm ground and mark as shown in the illustration
- protect benchmarks from site traffic and other dangers
- if an assumed/arbitrary height datum has been used for the scheme, for example as shown on contract drawings, check with the supervising authority that this datum may be used
- record position, reference number, level and date last checked of each TBM and the primary benchmark on the site plan(s)
- display updated copy of site plan or list of benchmarks (with details) in site office(s)
- check levels of TBMs at regular intervals†
- report any apparent disturbance of TBMs and immediately update TBM lists
- transfer levels from TBMs to permanent works as soon as practicable
- remove redundant TBMs from the site.

BS 5964: Part 2:1996 includes examples of setting-out stations and benchmarks, for short, medium and long-term use.

† *Warning:* Earthworks and ground settlement, heave, expansion or contraction will affect TBM levels.

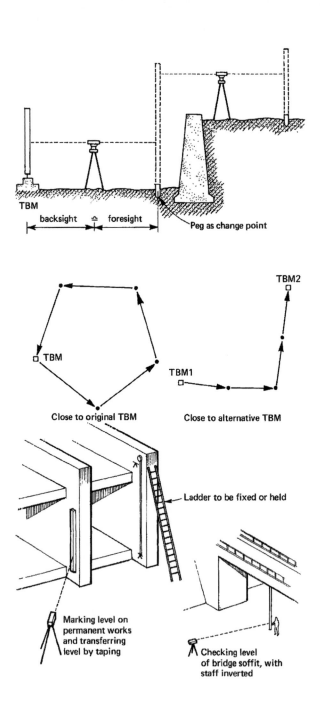

TBM

backsight ≙ foresight

Peg as change point

TBM

Close to original TBM

TBM2

TBM1

Close to alternative TBM

Ladder to be fixed or held

Marking level on
permanent works
and transferring
level by taping

Checking level
of bridge soffit, with
staff inverted

Levelling instruments

Whether the site engineer is measuring levels (eg for original ground levels or completed work) or providing levels (eg for blinding level for slab), the principles are similar.

Using and setting-up an optical level

- check collimation of instrument at least weekly (two-peg test)
- ensure firm base and comfortable telescope height
- avoid traffic and other vibration where possible
- arrange back-sight to be approximately equal to fore-sight
- do not leave instrument unattended.

Use of change points

Use peg, footplate or other convenient point to:

- avoid sighting more than, say, 40 m
- cater for large changes in ground level
- cope with obstructions.

Use of staff

- check that staff is undamaged
- ensure staff is held vertical or rocked (if no bubble is fitted)
- minimise use of extended staff. If extension used, check catch is fully engaged
- avoid sighting on bottom 0.5 m of staff (refraction is severe near ground)
- use inverted staff to measure soffit levels (book reading on staff as negative)
- having marked a given level (eg floor level) on a vertical surface (eg column face), the assistant should remove staff and reset staff to the mark before a check observation is made
- a tape is sometimes substituted for a staff (eg in a shaft).

Use of TBMs

- check for signs of disturbance and check reduced levels regularly
- transfer to permanent works as soon as practicable
- open and close levelling runs on different TBMs whenever possible
- check closing error is within acceptable limits.

Other levelling methods

- levels can be transferred by spirit level or water level
- a rotating laser can provide a horizontal reference plane but must be regularly checked against an optical level.

Use of precise optical level

- for establishing primary control on a large or extensive construction site
- for time-dependent deflections, eg pile load tests
- for precise levelling of mechanical equipment.

Use of digital levels

Digital levels use infrared or laser to take a reading from a bar-coded staff. When using digital levels:

- read the manufacturer's literature
- follow the same working practices as with traditional optical levels to reduce traversing errors
- ensure that the staff is always held vertical
- the staff must be in a well-lit area – self-illuminating staffs are available
- the infrared/laser may be affected by ambient conditions, eg heat haze, air temperature changes, which cause refraction errors.

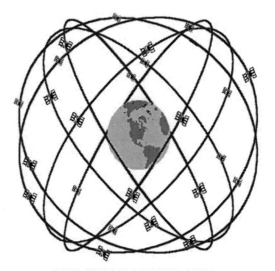

GPS SATELLITE CONSTELLATION

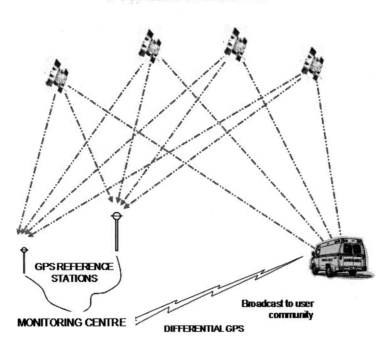

GPS REFERENCE
STATIONS

MONITORING CENTRE

Broadcast to user
community

DIFFERENTIAL GPS

Global Positioning System (GPS)

GPS (Global Positioning System) is a satellite-based positioning and navigation system owned and operated by the US Department of Defense. Access is free for all users and the service is available 24 hours a day, 365 days a year. GPS is an all-weather system that works anywhere in the world. GPS can give an instantaneous, three-dimensional real-time position to within approximately 10 m using a single channel handheld receiver. GPS is not the only Global Navigation Satellite System (GNSS). For example, the emerging European Galileo system will have satellites in orbit from 2008 and the Russian GLONASS system is also strengthening its constellation. Receivers that can pick up multiple constellations are available and bring the added advantages of being able to pick up more satellites (hence greater positioning availability) and accuracy.

The basic measurement of GPS is to determine the receiver – satellite range. The position of each satellite is known – this information is transmitted by all the satellites. The range can be calculated from either the code, which is modulated onto the transmitted signal, or the phase of the signal. If four or more satellites are used, a position for the receiver can then be computed.

Raw GPS positioning gives an accuracy of around 10 m. The accuracy can be improved by a technique known as differential GPS (or dGPS). This is where the errors in the GPS system (typically orbital, clock and atmospheric) are estimated and then used to correct the user's raw position. This can be done in one of two ways;

1   In real-time, through transmitting these errors to the user's receiver, for example via a phone, radio or the internet. The rover/mobile receiver uses these corrections to improve its position. A user can receive a correction from one of two sources:
    -   a commercial correction provider
    -   own base station set up on a known point.

2   In post processing, where the rover/mobile receiver's raw data is downloaded along with data from one or more base station receiver(s) set on known control point(s). Software is then used to correct the roving receiver's positions.

The accuracy of the end positions depends on the equipment and techniques used and the local conditions. For example, high quality permanent receivers can be positioned with a precision of a few millimetres with careful post-processing. An application for this could be in dam or bridge monitoring. Surveying grade equipment can give absolute accuracies from 1 cm in either real-time or post-processing. Lower grade equipment can then be used for other applications. Knowledge and understanding of the equipment, its use and the possible error sources is paramount when using GPS equipment and requires expert operators. The GNSS references in the bibliography section, or other texts, should be reviewed before embarking on a GPS survey.

For setting out operations, Real Time Kinematic (RTK) GPS/GNSS equipment is excellent for some operations – *see* manufacturers for the applications and accuracies that are available. Local grid or National Grid coordinates are entered into the receiver or handheld controller and then the unit tells the surveyor exactly which direction to go to locate the point to be set out.

GPS can be used in conjunction with Total Stations and traditional equipment to bring National Grid or Ordnance Datum heights into a site, plus to create control and reference objects for optical work.

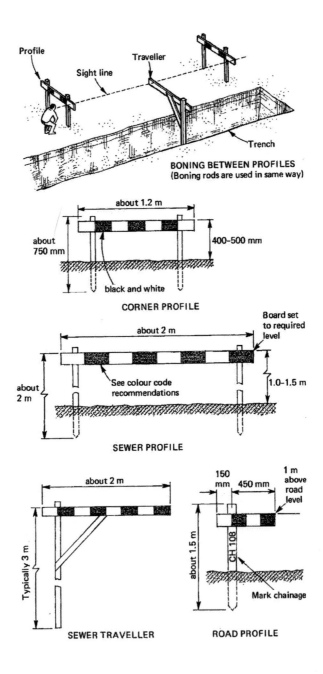

Profile

Sight line

Traveller

Trench

BONING BETWEEN PROFILES
(Boning rods are used in same way)

about 1.2 m

about 750 mm

400–500 mm

black and white

CORNER PROFILE

about 2 m

Board set to required level

See colour code recommendations

about 2 m

1.0–1.5 m

SEWER PROFILE

about 2 m

Typically 3 m

SEWER TRAVELLER

150 mm

450 mm

1 m above road level

about 1.5 m

CH 108

Mark chainage

ROAD PROFILE

Profiles and travellers

Standardisation

As far as is practicable, it is helpful to standardise on the construction and colour coding of all profiles and travellers used on a site. Appropriate sizes are provided in the illustration opposite. For a suggested colour scheme and other details, *see* Suggested Colour Code at end of this document.

Use of profiles and travellers

Profiles and travellers are used for providing lines and levels/grades. The general principle of 'boning' between profiles or boning rods is illustrated in the diagrams opposite.

Setting up profiles

- drive long pegs into the ground at suitable positions, for simple excavations (eg level formations), these locations are not critical. For pipelines or road formations the peg position will affect the value of the levels to be maintained and should be as close to the work as practicable
- determine the reduced level of the top of each peg relative to the nearest site TBM; check the level
- calculate the required level at each peg position. Add a suitable traveller height (the whole process depends on the assistant's eye, so the traveller height must not be too low or high for comfort)
- calculate the difference between the required level and the top of peg levels. If the peg is higher, mark this difference measuring vertically down from the top of the peg and fix a profile board at this height. If the peg is lower, make up a T-piece to raise the profile board by the required amount above the top of the peg
- check that the profile board is horizontal, adjust as necessary.

Sewer profiles and travellers

Sewer profiles are set at a given height above the required trench formation level and to one side of the trench. The horizontal board of the traveller is made to project to that side of the trench so that it can be 'boned through' the profiles. Ensure that the profile will not be obscured by spoil heaps etc.

Each traveller should be marked with its height together with the manholes between which it passes (eg MH 26–MH 30).

Corner profiles

A pair of profiles can be used to define a construction line or parallel lines (eg faces of foundation trench and outside faces of brickwork). The profile boards may also be set to level and used with a traveller to 'bone in' a required level (eg bottom of foundation trench).

The boards should be set outside of the supporting stakes so that a string-line stretched between the profiles will not pull the boards off.

Road profiles and travellers

Road profiles are usually provided in pairs, one each side of the centre-line, at intervals along the line of the road. The chainage at each profile should be marked on each profile. Batter rails are then set outside of the limits of the earthworks to control the bulk cut and fill operation. When the earthworks are complete, new profiles are set just clear of the road to control final trimming of the formation level and construction of the road. For the latter, a traveller on a base is helpful, with a facility for adjusting the height of the board to cater for the various component layers of the road.

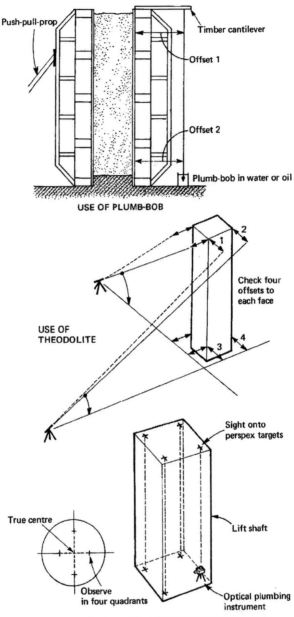

Push-pull-prop

Timber cantilever

Offset 1

Offset 2

Plumb-bob in water or oil

USE OF PLUMB-BOB

USE OF THEODOLITE

2
1
Check four offsets to each face
3
4

Sight onto perspex targets

Lift shaft

True centre

Observe in four quadrants

Optical plumbing instrument

USE OF OPTICAL PLUMBING INSTRUMENT

Establishing the vertical

Verticality to a modest accuracy can be achieved with a good quality 1 m builder's spirit level, but other methods must be used as the height and/or need for precision increases.

Use of plumb-bob (plummet)

In the example shown in the diagram, a wall is checked for verticality. The plumb-bob is suspended from a piece of timber nailed to the top of the formwork and shielded from the wind or immersed in a pail of oil or water to keep it steady. Offsets from the back of the form are measured at top and bottom taking account of any steps or tapers in the wall. Any necessary adjustments are made with a push-pull prop.

Use of theodolite or Total Station

In the example opposite, the formwork for a tall column is plumbed for verticality. A theodolite or Total Station is set up on a plane parallel, but offset, to one face and sighted on suitable offset marks at the top. (Observe both edges to check on twist). Similar observations are then made on the bottom of the formwork. Any discrepancy in verticality from an average of the observations is read at the bottom for convenience and the column form adjusted. The whole process is then repeated for the adjacent face.

Sighting at a steep angle is made easier by using a diagonal eyepiece.

Note: For accuracy, the instrument should be some distance from the column; this may be impossible in confined areas. A well-checked and collimated instrument is mandatory.

Use of optical plumbing instrument

The operation is relatively simple:

- set up and level the instrument over the ground station ensuring the instrument is correctly located above the ground marker
- sight up onto a target board and have the assistant mark the spot
- turn the instrument through 90°,180° and 270° in horizontal plane to define three further points, ie observe in four quadrants to remove any error in the instruments verticality
- the intersection of the diagonals joining the four points lies on the vertical above the ground point.

Optical plumbing is particularly useful for ensuring the accuracy of lift shafts, slip-formed structures and climbing forms. The example opposite shows the use of an optical plumbing instrument in a lift shaft, using Perspex targets fixed at the top level. At least three ground stations should be used to check for possible twisting in the shaft. A well-checked and collimated instrument is mandatory to avoid systematic error.

Use of lasers and auto-plummets

Laser plummets are similar to the optical plummet with the exception that a laser beam replaces the observer's eyeline. If properly adjusted and correctly used, laser plummets can be superior to optical plummets. In more advanced instruments, the verticality is maintained automatically through a compensation device. Nevertheless, the prudent engineer will still rotate the plummet through a full circle to ensure that any error in collimation is removed.

Verticality over long distances

Maintaining verticality in tall structures/high-rise buildings etc, is a specialised practice beyond the scope of this guide and a cautious engineer will always call in experts qualified in high-rise setting-out techniques.

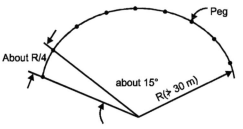

USING CENTRE POINT

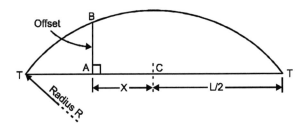

USING CHORD POINTS

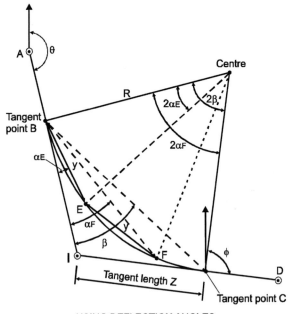

USING DEFLECTION ANGLES

Circular curves

Using the centre point

Circular curves of 30 m radius or less can be set out with a steel tape from a known centre point, placing pegs equidistant around the circumference of the circle at intervals of about 15° of arc or ¼ of the radius apart.

Using chord points

If the centre point cannot be used, a short curve radius can be set out by offsets, given two points on the circle and the radius, R:

- measure length of chord, L (if not given)
- set up string line between chord points, T
- set out centre of chord, C
- for distance X along chord from C, set out offset AB perpendicular to chord where:
  $$AB = \sqrt{(R^2 - X^2)} - \sqrt{(R^2 - (L/2)^2)}$$

Using deflection angles

Large radius horizontal circular curves can be set out by deflection angles with a theodolite or Total Station. In the example shown opposite, the coordinates of A, B, C, and D and the required radius R are given.

The procedure is:

- from the coordinates of A and B, calculate whole circle bearing $(WCB) = \theta$
- for first point on curve (E) choose convenient chord length Y
  [suggested chord intervals: retaining walls 2–5 m. kerbs 5–10 m, earthworks 20–30 m]
- calculate deflection angle $\alpha = \sin^{-1}(Y/2R)$
- calculate WCB of E from $B = \theta \pm \alpha$ (minus in the example)
- repeat for further points F etc, but note that for point C chord length is calculated from coordinates of B and C.

With instrument set up on B set out points on the curve. A worked example can be found in Appendix D: Road layout: traditional method Proving setting-out stations.

For a long curve, or where sight lines are obstructed, it will be necessary to set out the curve in sections, moving the instrument for each section.

Note: If intersection point I is given rather than coordinates of C:

- calculate tangent length Z from coordinates of B and I
- calculate deflection angle $\beta = \tan^{-1} Z/$
- calculate chord length $= 2R \sin \beta$.

Using a Total Station

Setting-out any curve, where the coordinates of the curve are provided (or can be calculated), is greatly simplified, quick and offers the potential for greater accuracy by using the coordinate setting-out functions of a Total Station.

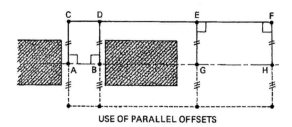

USE OF PARALLEL OFFSETS

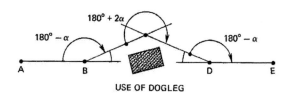

USE OF DOGLEG

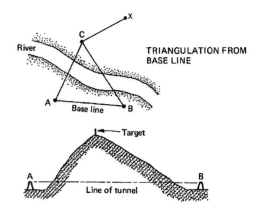

TRIANGULATION FROM BASE LINE

USE OF COORDINATES

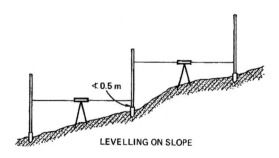

LEVELLING ON SLOPE

Obstructions and slopes

All possible problems caused by obstructions and slopes cannot be anticipated but the following hints may help.

Use of parallel offsets

To extend a line beyond an obstruction, a parallel line can be set out by offsets to get around the obstruction then return to the original line. This technique, while useful, demands utmost care in setting-out the offsets as a small error can cause a large discrepancy at the line terminal.

In the example, C and D are offset from A and B at right angles to AB. CD is extended to E and F beyond the obstruction. Finally, G and H are offset from E and F to bring the line back to the original direction.

Note: The potential error is magnified because AB is short relative to the length of the obstruction. To mitigate the error, another offset line should be set out on the opposite side of the obstruction as a check.

Use of dogleg traverse

An alternative method is to use a simple dogleg traverse, noting the same need for accuracy and checking:

- set up instrument on B, sight on A, then turn the instrument $(180° - \alpha)$ onto an intermediate point C. Measure distance BC
- at C, set up instrument and sight on B, then turn an angle $(180°+2\alpha)$ and set out distance CD equal to BC to establish point D
- at D, sight on C, turn angle $(180° - \alpha)$ and set out point E (and hence extend the original line).

Use of baseline

Inaccessible points can be set out from a known baseline by triangulation, for example across a river. In the example, two stations A and B provide the baseline:

Option A: via intermediate point

- set out a peg C close to where the setting-out point (X) is to be located
- with the instrument at A, observe the angle CBA
- with the instrument at B, observe the angle ABC
- with the instrument at C, observe the angle BCA
- sum the angle to ensure they add up to 180°
- calculate bearing and distance of X from C and set out as normal

Option B: direct setting-out

a) calculate bearing A to setting-out point C, B to setting-out point C, and also from A to B
b) at A, observe B and set bearing A to B on the instrument, turn instrument to bearing A to C
c) with the instrument locked, set out a marker at a short distance beyond point C
d) repeat step c) setting-out a point a short distance in front of point C
e) at B, repeat steps b) to d) substituting the bearings B to A and B to C
f) stretch string lines between the markers and where the lines cross marks point C
g) measure angle at C between B and A to ensure the triangle sums to 180°
h) check observations and adjust as necessary.

Levelling on slope

On a slope, keep the distance between back-sights and fore-sights reasonably equal and short to avoid introducing cumulative error. If levelling up the slope, ensure the line of collimation is not less than 0.5 m above ground level at the proposed peg location.

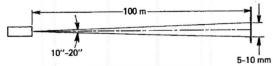

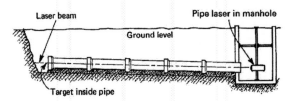

DIVERGENCE OF LASER BEAM

Laser beam

Pipe laser in manhole

Ground level

Target inside pipe

PIPE-LAYING

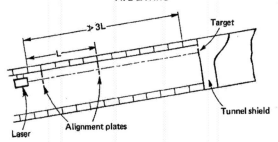

Target

Tunnel shield

Laser

Alignment plates

TUNNELLING

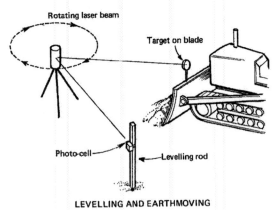

Rotating laser beam

Target on blade

Photo-cell

Levelling rod

LEVELLING AND EARTHMOVING

Lasers

Only a cursory overview of laser equipment is given as these are instruments that are developing rapidly.

Types of laser

Setting-out laser instruments include:

- alignment laser or rotating laser to define a plane
- visible beam (He–Ne) or invisible beam (eg gallium-arsenide)
- Class 1, 2, or 3A as defined by BS EN 60825

Safety and limitations

The following safety precautions should be adopted when using lasers:

- use Class 1 or Class 2 lasers to avoid eye damage from brief contact with the beam
- Class 3A lasers should be used only by competent and qualified personnel
- in all cases, avoid installing the laser at eye level.

The minor limitations of lasers are:

- beam divergence – typically 5 to 10 mm per 100 m
- beam refraction due to non-uniform temperatures or where it passes too close to solid material (eg a tunnel wall). Where possible, ventilate tunnels and ducts to achieve uniform temperature gradients
- invisible beam requires photoelectric detector.

Use of alignment laser

A visible beam is generally most convenient. For pipe laying, the laser is set to line and gradient for:

- excavation (using traveller with solid target)
- laying pipes (using transparent target inside pipe).

The procedure is usually:

- set out pegs for centre of chambers (using coordinated data where given)
- offset the pegs and reference. Excavate chamber to reduced dig level and place blinding concrete to invert
- excavate, lay and bed two pipe lengths using theodolite or a level
- set up laser in chamber invert/pipe invert to the required gradient
- carry out periodic checks to invert level (say every 10th pipe) to ensure laser is operating within tolerance.

For tunnelling, the procedure is normally:

- fix two vertical plates with small drill holes, offset from the tunnel centre, to the completed tunnel roof
- the laser is fixed to the tunnel roof and adjusted until the beam passes through the holes in the two plates
- the beam projects onto a target on the tunnelling shield, or tunnelling machine, for control of line and gradient
- to preserve accuracy, the laser is moved forward and reset when the target distance reaches maximum (see diagram opposite).

Use of rotating laser

Both visible beam and invisible beam rotating lasers are available but the latter avoids the potential nuisance of a visible occulting beam. All rotating lasers should be gyro-compensated and able to operate in the horizontal or vertical plane. Typical uses include:

- earth-moving to level on horizontal plane
- providing a horizontal reference plane across a site
- providing a vertical reference plane for curtain walling.

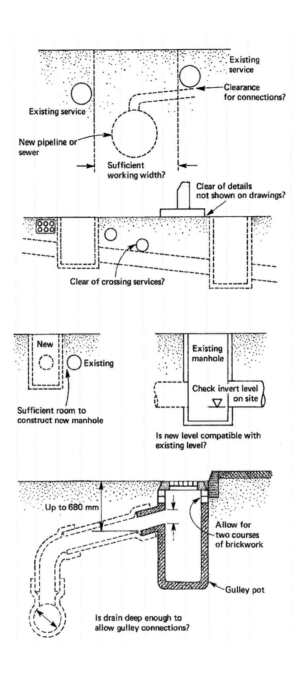

Existing service

Clearance for connections?

Existing service

New pipeline or sewer

Sufficient working width?

Clear of details not shown on drawings?

Clear of crossing services?

New

Existing

Sufficient room to construct new manhole

Existing manhole

Check invert level on site

Is new level compatible with existing level?

Up to 680 mm

Allow for two courses of brickwork

Gulley pot

Is drain deep enough to allow gulley connections?

Sewers and drains: initial checks

Before setting-out sewers and drains, the site engineer should mark the approximate line(s) then:

- walk the line, noting existing features and correlate with the site drawings. note and report any variations
- note any potential setting-out or construction difficulties and consider solutions
- where possible, check soil conditions
- clear any discrepancies and agree the proposed method with the supervising authority before proceeding with the detailed setting-out.

Adjacent and crossing existing services

The supervising authority should have notified the statutory undertakers of the proposed works, obtained details of adjacent services and checked these thoroughly before proceeding to ensure adjacent cables and pipes will not be at risk. The site setting-out engineer should:

- consult public utilities drawings of mains and cables to ensure that proposed sewers and manholes do not conflict with these services. If in doubt, dig trial holes
- do not assume that an existing sewer follows a uniform gradient or straight line between manholes. Do not forget to take into account pipe thickness and collars or patent joints
- mark up drawings with all available information and confirm these with the statutory undertakers
- check that the distance between new and existing services will be sufficient for excavation, trench support and working space, including construction of manholes.

Warning: Record drawings of existing services are frequently unreliable.

Interference by drain connections

Inspect the drawings for the location of drainage connections and determine whether these might interfere with any adjacent services, particularly as regards to level/depth.

Note: House services are unlikely to be shown.

Manhole positions

Agree and confirm with the supervising authority the intended location of the manholes on the ground. Normally, sewer drawings do not show exact positions or take account of local details.

Discharge levels

If a new sewer is to discharge into an existing manhole or outfall, check physically on site that the level shown on the drawings is accurate. This is vital – errors may require redesign of part or whole of the system.

Gully connections

Check that the surface water sewer invert level is sufficient to accommodate connections from road gullies.

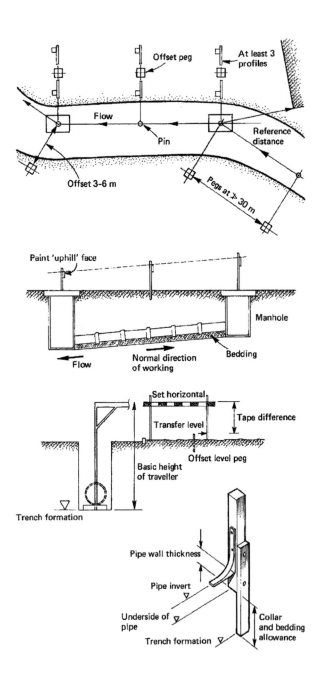

Offset peg
At least 3 profiles
Flow
Pin
Reference distance
Offset 3-6 m
Pegs at ≯ 30 m

Paint 'uphill' face
Manhole
Flow
Normal direction of working
Bedding

Set horizontal
Tape difference
Transfer level
Offset level peg
Basic height of traveller
Trench formation

Pipe wall thickness
Pipe invert
Underside of pipe
Trench formation
Collar and bedding allowance

56

Sewers and drains: line and level

Centre-line

For line and level, set out the centre-line of the trench with pegs or road pins typically <30 m apart. Mark the centre of each manhole with a peg, referenced to other pegs or nearby permanent features.

Offset pegs

Typically, offset pegs or road pins are positioned 3 to 6 m to one side of the centre-line. The offset distance should take account of where excavated material will be placed and the space required for the free movement of plant. The site engineer should strive to keep offset distances between manholes constant.

Profiles

Set up standard profiles on the same side of the centre-line as the offset pegs. If possible, the end of the profile board should be not more than 3 m from the centre-line. Paint 'uphill' face of profile board (see diagram opposite).

Find level of offset peg and calculate depth to the pipe invert or the trench formation level before deciding the height of the traveller. Calculate the distance from the offset peg to the top of the profile board and fix board by transferring peg level to stake and measuring upwards as shown. Indicate on profile its height above invert level or trench formation as appropriate.

Wherever possible, avoid using a traveller less than 2 m high. Provide one profile at each manhole and at least one between manholes. The middle profile(s) allows a quick check on the accuracy of the other two. Calculate the level(s) for the middle profile(s) from the levels and gradient shown on the drawings.

Travellers

The basic height of the traveller should be the difference in level between the line defined by the profiles and the pipe invert, or the trench formation level, whichever is appropriate, and should be a convenient multiple of 0.5 m. Excavation level, bedding level and invert level can all be indicated, if wished, as in the diagram opposite. Mark the traveller with its height and the manholes between which it is to be used.

Information sheet

Provide foremen/gangers with Site Information Sheets recording relevant information on pegs, profiles, travellers and bedding.

Suggested colour code

A suggested colour code for pegs and profiles is shown on the end papers.

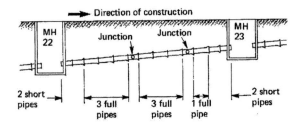

LOCATION OF JUNCTIONS

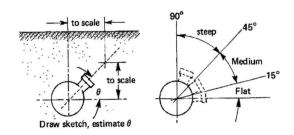

ANGLE OF JUNCTION IN SECTION

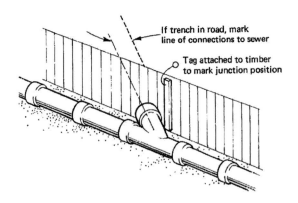

MARKING BEFORE BACKFILLING

Sewers and drains: junctions

Establishing location

When setting-out junctions for sewers and drains, the normal procedure is:

- show the plan location of each junction on a Sewer Information Sheet (and Site Information Sheet as appropriate)
- express the location of each junction in terms of the numbers of pipes to be placed from a given manhole before the junction, or the number between junctions
- measure up the position for later fitting of saddle connections.

Note: Specifications often require that the pipe built into the manhole and the next pipe be short pipes to avoid excessive bending load on the pipe. In the example opposite, the first junction is placed after five pipes (two short and three full) and there are three full pipes to the second junction.

Estimating angle

- estimate angle of junction in plan and specify type of junction required on information sheet
- estimate the angle of each junction in section and classify as flat, medium or steep as shown
- provide foreman and ganger with copy of the information sheet to begin work.

Note: If angles are difficult to determine, seek the supervising authority's agreement to use saddle connections.

Marking before backfilling

- mark junctions before backfilling for ease of constructing trenches and for making connections
- measure the 'as laid' location of each junction by taping from the inside face of the nearest downstream manhole
- record position on 'master copy' of Sewer Information Sheet.

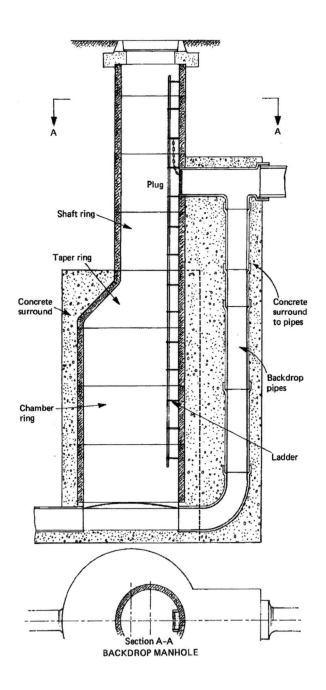

Plug

Shaft ring

Taper ring

Concrete
surround

Chamber
ring

Concrete
surround
to pipes

Backdrop
pipes

Ladder

A

A

Section A–A
BACKDROP MANHOLE

Sewers and drains: manholes

Manholes in general

The site engineer should prepare a Site Information Sheet containing all the relevant information for the foreman/ ganger.

The following should be included as appropriate:

- base location, dimensions and level
- type and number of concrete rings (if used)
- taper
- invert level
- channel and benching details
- concrete surround (if required) cover slab level
- location of man-entry hole and step-irons or ladder
- any divergences from typical manhole details.

If concrete rings are used:

- designate each specific type by a code letter (to correspond with information sheet)
- paint code letters on rings on delivery to site to ensure correct order of construction and overall depth.

Backdrop manholes

When dealing with backdrops in a sewer run, errors can arise due to the abrupt changes in depth. To avoid errors, adopt the following procedure:

- draw attention to manhole on the sewer information sheet
- set up the profile posts for the section containing the higher run of sewer, but do not fix the profile board until the preceding downstream section is excavated
- when the higher section is ready for excavation, withdraw the traveller, re-set the profile and issue the new traveller. Details must be shown on a Site Information Sheet
- provide details of the construction of the backdrop and manhole to the foreman/ganger on a Site Information Sheet.

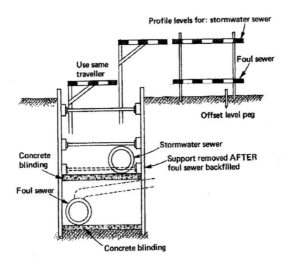

COMMON TRENCH

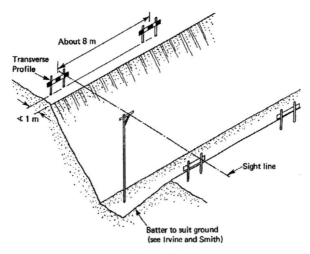

DEEP TRENCH (especially with battered sides)

Common trench

It is common for foul water and storm water sewers to be laid at different levels in a common trench.

Before setting-out, check that the invert level of upper sewer will allow connections to the lower sewer. To set out the sewers:

- prepare and issue a Site Information Sheet to the foreman/ganger
- erect upper and lower profiles as shown and paint uphill face of the sight rails (*see* end papers for colour code)
- ensure that trench sheeting will not obstruct boning on lower sight rail
- construct the traveller that will be used in conjunction with the upper and lower sight rail with one cross piece.

Deep trench

In a deep sewer trench with battered sides, if the usual profiles cannot be set within 3 m of the centre-line, and accurate boning is not possible, the following is recommended:

- set up pairs of profiles parallel to and on both sides of the trench at about 8 m intervals as shown in the diagram
- bone in traveller across the trench to give appropriate excavation level, leaving the ground a little high
- set up new profiles perpendicular to the trench line and within the trench for final trimming to level
- set up temporary datum level within the trench to set pipes to level.

In a deep supported trench, it is difficult to obtain accurate levels when the traveller is more than 5 m high, therefore:

- excavate roughly to depth using normal profiles
- set up profiles in trench and trim to final level.

Note: Beware disturbance of profiles in trench if trench sheets or sheet piles are being driven progressively. Check after each round of driving.

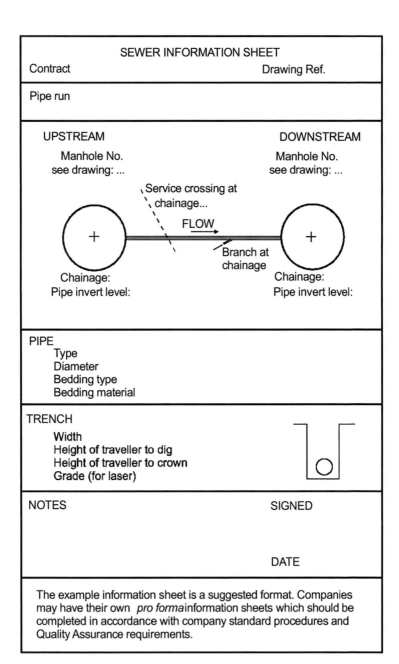

SEWER INFORMATION SHEET

Contract   Drawing Ref.

Pipe run

UPSTREAM   DOWNSTREAM

Manhole No.   Manhole No.
see drawing: ...   see drawing: ...

Service crossing at
chainage...

FLOW

Branch at
chainage

Chainage: Chainage:
Pipe invert level: Pipe invert level:

PIPE
    Type
    Diameter
    Bedding type
    Bedding material

TRENCH
    Width
    Height of traveller to dig
    Height of traveller to crown
    Grade (for laser)

NOTES   SIGNED

                                           DATE

The example information sheet is a suggested format. Companies
may have their own *pro forma* information sheets which should be
completed in accordance with company standard procedures and
Quality Assurance requirements.

Sewers and drains: Information sheets

Information sheets include hardcopy (paper) as well as computer/electronic documentation.

Sewer Information Sheet

The purpose of the Sewer Information Sheet is to ensure that site engineers and general foremen concerned with the supervision of the work are fully acquainted with what is required. The sheet also forms a record for future valuation purposes. The example sheet (*see* opposite) shows the minimum information required. Note that the direction of flow should be fixed (left to right). The manhole numbers should be inserted accordingly and the direction of working; this is usually in the direction opposite to the flow. (It is difficult to lay pipes downhill, especially those with collars).

Other information needed for guidance and record purposes should be added by the site engineer, such as:

- distances of offset pegs (alternative pegs provided on left and right of centre-line)
- known existing crossing services (marked in red)
- distance from downstream manhole to existing services or other potential obstructions across the line of the trench (to nearest 0.5 m or better)
- approximate lines of existing services parallel or nearly parallel and close to the line of the new sewer
- for junctions: diameter, location (number of pipes from downstream MH), direction (draw short line to left or right of centre-line) and angle (flat, medium or steep).

Site information sheet

The purpose of a Site Information Sheet is to provide the foreman/ganger with sufficient information for the job in hand. The information can be extracted from the relevant Sewer Information Sheet and any necessary notes added, for example, relating to trench support.

Sketches should be drawn on a Site Information Sheet so that a record can be retained in the site office.

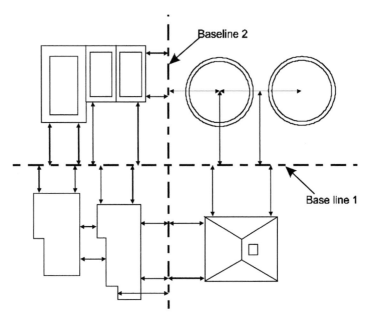

DIAGRAM OF SEWAGE DISPOSAL WORKS

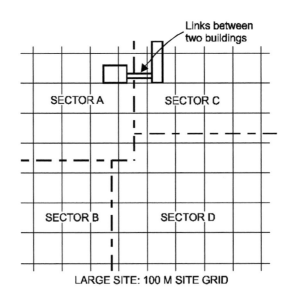

LARGE SITE: 100 M SITE GRID

Buildings and structures: location

Buildings and structures on same site

Where a project comprises several buildings or structures on the same site, the site engineer must ensure that individual units relate to each other so that the whole site is laid out in accordance with the site drawings. This is particularly important when pipe works or services connect one with another. It is vital that structures are set out to the tolerances specified.

For a small site, a simple system of baselines and offsets can be used as shown in the diagram for a sewage disposal works opposite. Two baselines at right angles are sufficient in this case.

Large site

A large site with a number of related structures poses additional problems that are mainly organisational. The site engineer may be one of a team, each engineer being responsible for the setting-out and control of a structure or part of a structure. Under these circumstances, it is essential for one engineer to coordinate the overall setting-out activity.

On a large site, there is no substitute for a grid of the whole site with a series of reference stations established round the site so that every part of a structure can be defined by coordinates and/or bearings and distances from the stations.

Each site engineer should:

- have a layout drawing of the whole site, showing the related structures, grid lines and main setting-out stations
- prove the main setting-out stations to be used
- check the setting-out by reference to structures or points already set out on adjacent sectors of the site
- liaise closely with the other site engineers for adjacent sections where structures are to be set out by direct reference to one another, both for location and level
- report immediately any apparent discrepancies in the site grid or between adjacent sections and resolve the conflicts.

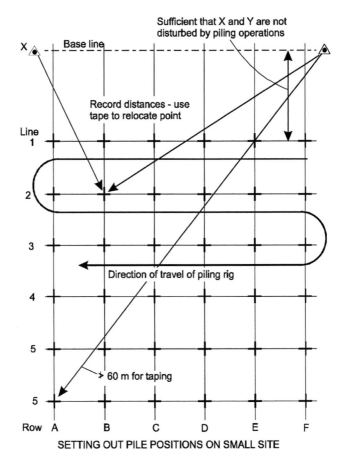

SETTING OUT PILE POSITIONS ON SMALL SITE

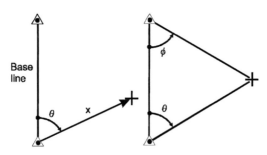

USE OF BEARING AND DISTANCE OR
INTERSECTION ON LARGE SITE

Buildings and structures: piling

Responsibilities

Piling may be sub-contracted to a specialist piling contractor. The piling contractor may then ask the main contractor to do the setting-out or may prefer to use his own engineer. In either case, the final responsibility for correctly setting-out normally remains with the main contractor's engineer.

Major problems associated with setting-out piles are:

- soil disturbance from driving displacement piles
- instruments disturbed by vibrations from piling process
- spoil from borings may obstruct fields of view
- piling rigs need space to manoeuvre
- most piles have to be set out individually
- the site engineer must constantly be available to assist and resolve difficulties.

Setting-out on small/medium sites

Where the number of piles is modest:

- set markers at the required positions, using a bricklayer's line and steel tape, from the corner profiles or site grid
- define a suitable baseline and record distances to each marker from at least two stations
- remove the markers, then
- relocate each pile position when the piling rig is ready to work.

Setting-out on large site

When the site is large and the pile positions spread widely, it may be more appropriate to locate pile positions by bearing and distance or by coordinates using a Total Station. Check critical pile positions, eg at the ends of a row of piles, by making redundant observations.

Levels

For cast-in-place concrete piles:

- install temporary level peg adjacent to the completed bore
- check founding level by using weighted steel tape
- give ganger dimensions from peg level or top of casing to concreting level (not cut-off level).

After installation

Before piling rig is removed from site:

- check construction tolerances of piles
- report any discrepancies on the pile log sheet
- agree remedial action with the supervising authority.

Raking piles

When setting-out raking piles it may be necessary to allow for the difference between the existing (working) ground level and the required level of the pile cap or pile cut-off level. Because of the inclination of the pile, the position at ground level will be displaced along the direction of rake by a horizontal distance $d \tan \alpha$, where d is the difference between the existing ground level and the specified pile cap or cut-off level and α is the inclination of the pile from the vertical.

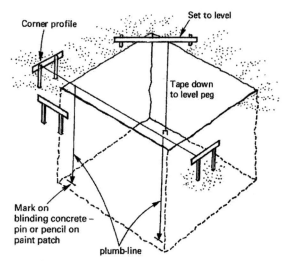

Corner profile

Set to level

Tape down
to level peg

Mark on
blinding concrete –
pin or pencil on
paint patch

plumb-line

NOTE: Support to excavation omitted for clarity
**TRANSFERRING LINES AND LEVELS TO BOTTOM
OF EXCAVATION**

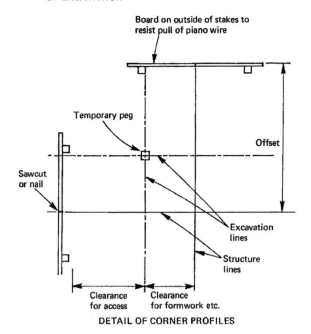

Board on outside of stakes to
resist pull of piano wire

Temporary peg

Offset

Sawcut
or nail

Excavation
lines

Structure
lines

Clearance
for access

Clearance
for formwork etc.

DETAIL OF CORNER PROFILES

If topsoil is to be removed before setting out the excavation, the site engineer should set out in two stages.

Stage 1: Removal of topsoil

- determine overall excavation dimensions
- set out temporary pegs defining area for excavation – sprinkle sand along string line between pegs as guide.

Stage 2: Main excavation

Once the area has had the soil removed, proceed as follows (refer to diagram opposite for clarification):

- set up stakes for corner profiles offset from structure lines by a convenient multiple of, say, 1 m
- set and level peg adjacent to each profile
- set profile boards to a level such that the height of traveller for excavation is a convenient multiple of, say, 0.25 m (traveller preferably not more than 5 m high)
- check that clearance between profiles and excavation lines is sufficient for plant access
- set out structure lines by marking opposing profiles
- set out excavation lines from corner lines by distance needed to give clearance for formwork etc
- provide foreman/ganger with dimensioned sketch on a Site Information Sheet showing profiles, excavation lines, structure lines and traveller height.

Structure lines

Shallow excavations:

- use theodolite or Total Station if sufficient formation is visible.

Deep excavations

- set up piano wire between opposing profiles
- plumb down line as close to corners as practicable.

Levels

Excavations <5m deep

- bone between profiles to give four edge strips to level
- use boning rods to level dumpling
- set pegs for blinding by levelling from TBM.

Excavations >5m deep

- set up board across corner of excavation to a reduced level
- tape down to level peg at bottom of excavation
- set up temporary profiles within excavation
- set pegs for blinding by re-taping
- agree peg levels with the supervising authority or seek other independent check.

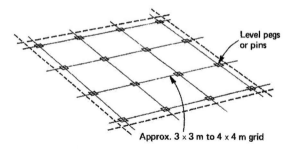

PEGS FOR BLINDING CONCRETE

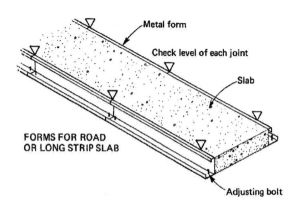

FORMS FOR ROAD
OR LONG STRIP SLAB

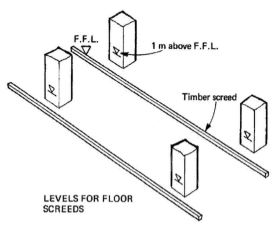

LEVELS FOR FLOOR
SCREEDS

Buildings and structures: slabs & floors

Ground slabs

To set out ground slabs, the site engineer should:

- set up profiles to control excavation to formation level
- drive 300 mm steel pins to level on a grid of between 3 and 4 m centres to control blinding.

If ground slab is to be screeded, the edge form controls the structural thickness of the slab. Minor variations in level are absorbed within the screed thickness. If no screed is required, set timber or metal forms to level to control finished floor level. Metal forms are preferable for accurate results and frequently incorporate adjusting bolts. Check level at each joint between forms.

More careful levelling may be necessary for warehouses or other buildings where mobile plant, eg fork life trucks, will operate over the floors.

Suspended slabs

Setting-out levels for falsework:

- determine the required soffit levels from site drawings. Make any necessary allowances for pre-camber or 'set' of formwork. For simple (flat slab) floors these levels will all be the same; in complex cases (eg curved or super elevated bridge decks) each position will have a unique level
- determine falsework (forkhead) levels allowing for depth of materials, primary and secondary beams plywood decking etc, and any deflection under load (eg in long spans)
- calculate the approximate scaffold heights required and then top and bottom jack lengths
- mark check levels on the adjacent structure where possible, eg by pencil or chalk line on columns, walls or abutments. It is useful to mark both soffit level and an offset line say 500 mm below, but make it clear which is which
- level the forkheads once erection of the scaffold is sufficiently advanced. If levelling from the ground, hang a tape measure from the forkhead rather than use an inverted staff
- adjust forkheads as necessary – it is easier at this stage, before the falsework is loaded
- when the formwork (timbers, beams and soffit) is complete, but before reinforcement is fixed, check levels at the forkhead positions and adjust if necessary. To do this, the scaffold standard layout may need to be marked for reference on the soffit
- if the slab is heavily reinforced, the soffit may need to be set slightly high, say 5 mm, to allow for additional settlement in the soffit materials
- rebar (reinforcing bar) levels may need checking on deep sections to ensure cover is correct.

Floor screeds

Mark columns 1 m above finished floor level. The screed layers will set timber screeds from these levels. Carry out spot checks on timber screeds before screed is laid.

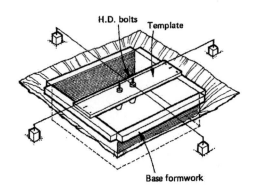

STANCHION BASE

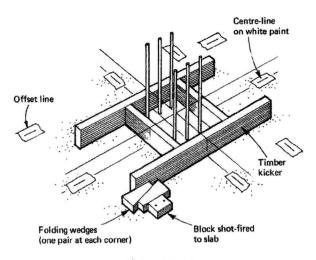

COLUMN KICKER

Stanchion or column bases

To set out the bases of columns and stanchions, proceed as follows:

- using a theodolite or Total Station, set out the centre-lines of the stanchions or columns with four pegs outside the area excavated for the base
- when the base formwork has been fixed in relation to these pegs, re-set the instrument, check pegs and mark centre-lines on plywood or timber template nailed across top of form. Holding down bolts, mortise boxes or column kickers are set out from centre-lines by carpenters
- level from TBM to mark concreting level on form
- check that formwork or template has not been disturbed during placing of concrete (use pegs)
- reset the instrument and make a final check on template position after concrete is placed but before final set.

Column and wall kickers

The centre-lines of columns or walls, together with any offset lines, should be set out on a previously cast slab so that the carpenters can locate the kicker form. It is imperative to set out these lines as soon as the slab has hardened and before it is used for stacking reinforcement, formwork and other materials. Ensure that the marked lines will remain accessible.

In the diagram opposite, note the use of timber blocks shot-fired to the slab to hold the form in position with folding wedges.

Column and wall formwork

- check that the kickers are aligned with the marked centre-lines and are square in both plan and elevation
- check formwork for verticality with plumb line or theodolite immediately before placing concrete
- provide concreting level from local TBM, taping vertically or levelling as necessary, mark with a chalk line, nail through shutter or batten or angle fillet to suit finish to top of wall. In some circumstances (eg where the wall will be extended upwards), a dipstick may be adequate to define concrete level
- check verticality of formwork after concrete is placed but before final set.

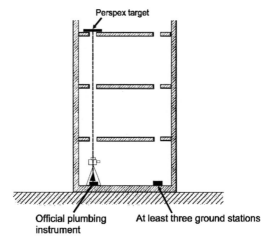

Perspex target

Official plumbing instrument

At least three ground stations

PLUMBING MULTI-STOREY BUILDING

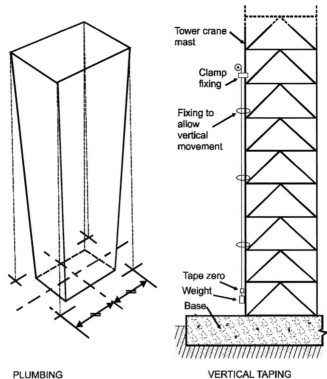

Tower crane mast

Clamp fixing

Fixing to allow vertical movement

Tape zero

Weight

Base

PLUMBING
TAPERED COLUMN

VERTICAL TAPING

Buildings and structures: tall structures

The following guidelines are appropriate for most tall structures less than 15–20 m high. For taller or more complex structures, expert guidance should be sought.

Plumbing

The primary instruments for plumbing tall structures[1] are:

- plumb-bob
- theodolite or Total Station
- optical or laser plumbing instruments.

The principles of use are described in section Establishing the vertical

Verticality and twist

When setting-out tall structures, checking for verticality and twist must be monitored frequently. This is best achieved by plumbing up or down from four fixed points.

Rectangular structures

The use of an optical plumbing instrument to control the verticality of a multi-storey building is illustrated in the diagram opposite.

The plumbing of lift shafts is a particularly critical operation because the installation tolerances are small. Four setting-out points should be established at the base of the lift shaft such that the vertical lines through them will not be obstructed by formwork or scaffolding. Plumbing can be from top to bottom using plumb-bobs or from bottom to top using an optical/laser plumbing instrument, particularly in conditions where wind or draught might disturb plumb-bob line.

Tapered structures

To plumb a tapered column or similar structure:

- set out two centre-lines on the base at right angles
- plumb from top corners (plumb-bob or theodolite) and check equality of offsets on all four sides (*see* diagram).

If the structure narrows towards the top, it will be necessary to cantilever out from the top of the formwork to fix plumb-bobs.

Height and level

Floor-to-floor dimensions can be controlled using a weighted steel tape, measuring each time from a datum at the base of the structure. Each floor is then provided with datum marks in key positions from which to transfer levels on each floor. Note that the weight should be kept constant (and preferable equal to the tension under which the steel tape is in calibration) and allow for temperature variations.

The base datum level should be set in a location that allows unrestricted taping to roof level. If a tower crane is used, a tape can conveniently be fixed to the mast (*see* diagram).

Warning: Errors, apparent or real, can result from differences in thermal movement of the tape relative to that of the building, particularly if construction spans a number of seasons.

---

[1]*Caution*: for high rise structures, these methods are inadequate and expert advice should be sought.

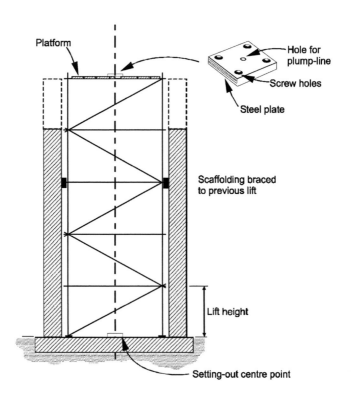

Platform

Hole for plump-line

Screw holes

Steel plate

Scaffolding braced to previous lift

Lift height

Setting-out centre point

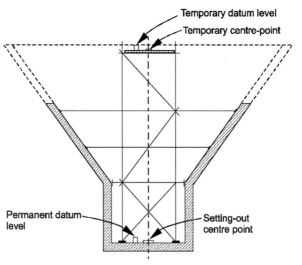

Temporary datum level

Temporary centre-point

Permanent datum level

Setting-out centre point

Constant radius

The simplest circular structure is where one or both of the internal and external diameters are constant. Setting-out such a concrete structure is straightforward:

- establish the setting-out centre point on the base
- set out kicker by taping and place kicker in position
- erect and plumb formwork and place first lift
- strip formwork
- erect scaffold tower braced against first lift of concrete with top platform level with top of second lift
- establish temporary centre point on platform by plumbing up or down from setting-out point
- set up second lift of formwork
- check formwork radius from temporary centre point
- repeat sequence for further lifts.

Varying radius

The same basic procedure as above is followed except that it is necessary to calculate the radius at a given height and to control that height. The sequence of operations is amended as follows:

- erect scaffold (as above)
- establish temporary setting-out point (as above)
- establish temporary datum level on top scaffold platform by taping from permanent datum level
- set up next lift of formwork to level
- check radius of top of formwork.

Establishing temporary centre points

If a plumb-bob is used, suspend the bob from a line or piano wire passed through a steel or plywood plate. The plate can be screwed to the top platform and the plumb-bob adjusted until it lies centred over the setting-out point.

If an optical plumbing instrument is employed, sight upwards to a Perspex target fixed to the top platform and mark as before.

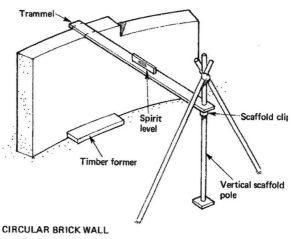

CIRCULAR BRICK WALL

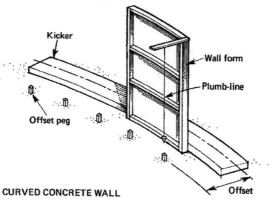

CURVED CONCRETE WALL

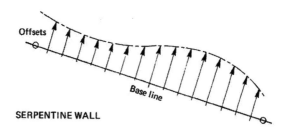

SERPENTINE WALL

Circular wall – centre accessible

If the centre is accessible and the radius not greater than, say, 30 m, it is convenient to set out the first course of a brick wall by taping from the centre (similarly for the kicker of a concrete wall). The bricklayer works by plumbing vertically with a spirit level and checking segments of the curve with a timber shaped to the radius.

If the centre is available, and the radius is less than say, 5 m, a scaffold pole can be set up as the centre for a trammel board as shown in the diagram. The inner and/or outer radius is marked on the trammel, which is raised course-by-course with a scaffold clip to control the level of the wall.

Circular wall – centre not accessible

Where the centre is not accessible, the curves must be set out using deflection angles and chord lengths. Chord intervals should be about 3 m.

For example, the method for a concrete wall could be:

- set out centre-line
- establish offset pegs
- erect profiles, excavate and place foundation
- set out kicker accurately from offset pegs – place kicker
- set up forms and plumb down to offset pegs or secondary offset points.

Serpentine walls

Serpentine or S-shaped walls are difficult to set out accurately by using deflection angles and chord lengths. An alternative is to use a baseline and offsets, so long as the offsets lie within a convenient taping distance. Offset pegs can be established on the baseline at suitable intervals. The offset distances can be calculated or taken off the drawings, provided the geometry of the wall is known.

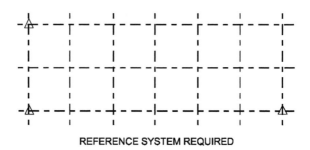

REFERENCE SYSTEM REQUIRED

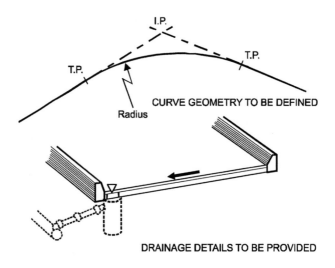

CURVE GEOMETRY TO BE DEFINED

DRAINAGE DETAILS TO BE PROVIDED

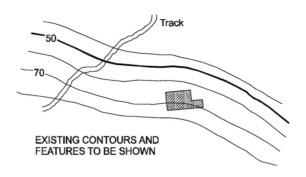

EXISTING CONTOURS AND
FEATURES TO BE SHOWN

Roads: checking drawings

Before setting-out roads, the site engineer should mark the approximate route(s) then:

- review the drawings together with some of the general points to be checked
- walk the route, noting existing features and correlate with the site drawings. Note and report any variations
- note any potential setting-out or construction difficulties and consider solutions
- clear any discrepancies and agree the proposed method with the supervising authority before proceeding with the detailed setting-out.

Information to be checked

It is more likely that information may be less detailed for estate and similar minor roads than for major roads. Nevertheless, apply the checks to all drawings:

- where Ordnance Survey[2] National Grid coordinates and/or Ordnance Survey heighting/benchmarks are specified, *see* Appendix C: UK National Grid, benchmarks and ground distances for guidance
- reference points, a baseline, or a site grid, should be provided. If not, a suitable setting-out reference system must be agreed with the supervising authority
- details of existing features should be available. Check these against the latest Ordnance Survey large scale (1/1250 or 1/2500) plan and by walking the site
- existing contours or a grid of existing ground levels should be shown on the site plan
- check that fence lines exist and are accurately shown
- check that overall width between fence lines is sufficient to construct the earthworks
- borrow pits and tipping areas should be defined or agreed with the supervising authority. Check that these are adequate for the volume of earth to be removed or dumped
- access points to site, borrow pits and tipping areas should be defined
- horizontal curves should be fully defined with tangent points, etc. If not, design and agree suitable curve with supervising authority
- gradients, vertical curves, crossfalls or cambers and levels at junctions should be defined or agreed
- road drainage details should be shown. Check that there are adequate falls to gullies and along drains. For connections into existing drains, inverts should be stated and these checked on site.

Most often, information is presented in a computer file, *see* Appendix E: MOSS data for roads and Appendix F: Use of computers.

[2]Ordnance Survey triangulation pillars and marks together with benchmarks are redundant and are no longer maintained or supported. As the accuracy and reliability of these points is in doubt, their use is discouraged.

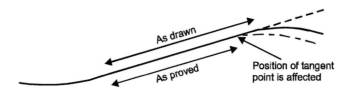

DISCREPANCY IN LENGTH

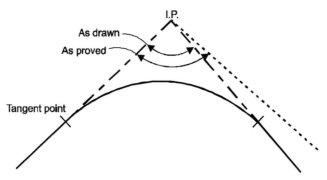

DISCREPANCY IN INTERSECTION ANGLE

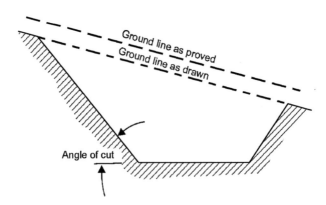

NOTE: Volume and width of cut are affected
DISCREPANCY IN EXISTING GROUND LEVELS

Roads: general procedures

Proving the survey

For major road schemes, the route survey and establishment of the road centre-line may be carried out by specialist surveyors. In this case, it should be necessary only to ensure that all the primary lines and levels have been established and that existing features agree with those indicated on the drawings.

For other road works, it is essential to prove the survey against the drawings before starting any setting-out:

- prove main setting-out stations (*see* Appendix D: Road layout: traditional method Proving setting-out stations)
- check, or establish, and agree primary and temporary bench marks
- establish intersection points on curves
- measure intersection angles
- measure tangent straights and chainages
- check position and coordinates of intersection and tangent points by reference to main setting-out stations
- check chainages at natural features, eg hedges and the relative positions of other features
- check existing ground levels along centre-line at intervals (use a suitable grid for wide roads or on steep side slopes)
- check if any existing feature or temporary works may obstruct the setting-out process.

Dealing with discrepancies

Discrepancies typically affect:

- road lengths
- intersection angles/design curves
- existing ground levels.

To minimise delay:

- double check that there is a discrepancy
- report the discrepancy and put forward proposed solution(s)
- agree action with supervising authority.

Note: Corrections to take account of discrepancies frequently require significant changes to land-take, gradients, cross-falls and drainage details.

Checking computer printout/files

If computer-derived setting-out information is provided, check that the changes in coordinates appear logical and that the ends of the section of road tie in with known features. *See* Appendix E.

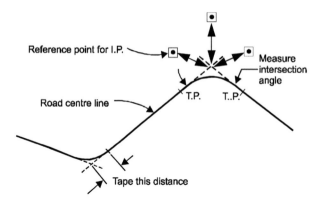

ESTABLISHING MAIN POINTS ON CENTRE-LINE

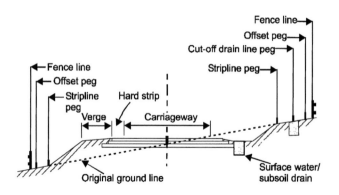

ESTABLISHING LINES BEFORE EARTHWORKS
(Typical rural road)

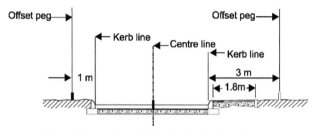

ESTABLISHING KERB LINES AFTER EARTHWORKS
(Minor access road)

Roads: line chronology

Establishing the centre-line

Normal practice is to supply road setting-out information as coordinates. These can be set out directly using a Total Station.

In the event that road design information is provided as traditional intersection points, tangent points etc, the conventional setting-out procedure begins with the centre-line as follows (*see* diagram opposite for clarification):

- establish intersection points from the site grid/reference station network
- set out offset reference pegs to intersection points
- measure intersection angles
- establish tangent points of curves by taping from intersection points
- set out curves.

An alternative to setting-out the centre-line is to set out the fence line accurately with a Total Station setting pegs every 20 m or so. The fence pegs will be outside the zone of earthworks and can be used to measure from for setting-out batter rails, first stage profiles, drains and ditches. Offsetting errors will be minimised.

Other lines and offsets

From the centre-line set out:

- fence lines
- drains and ditches
- site stripping area
- main chainage offset pegs (beyond limits of earthworks but within fence lines).

Use standard offsets where possible. Record all relevant dimensions on Site Information Sheet.

After earthworks complete

- re-establish centre-line from offsets
- check intersection and tangent points with respect to main setting-out points
- set out kerb lines
- set out offset pegs convenient distance beyond kerb line or footway.

Widening existing roads

Use the road surface for setting-out the centreline where appropriate. Road nails or shot-fired nails should be used to define the permanent line from which to set out offset pegs.

Pins should be numbered in white paint on the road surface and the following shown on an information sheet:

- pin number
- chainage
- offset distance to point 1 m beyond face of new kerb.

Provide separate standard level pegs or pins.

Suggested colour code for pegs

A suggested colour code for pegs (and profiles) is shown on the end-papers.

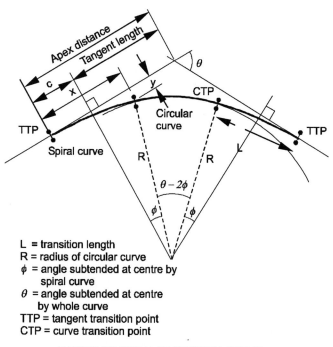

L = transition length
R = radius of circular curve
ϕ = angle subtended at centre by spiral curve
θ = angle subtended at centre by whole curve
TTP = tangent transition point
CTP = curve transition point

BASICS OF SPIRAL TRANSITION CURVE

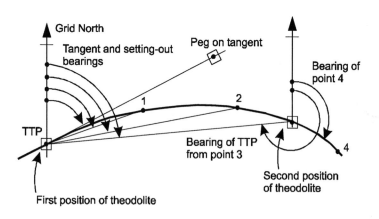

SETTING OUT CURVE

Roads: horizontal curves

Circular curves

Many minor roads incorporate simple circular curves only. These may be set as described previously, by using deflection angles or, where presented in coordinate form, by Total Station.

Major roads designed for high-speed traffic employ spiral transition curves, *see* next.

Specification of spiral transition curves

It is usual for road information to be presented in coordinate form for setting out with a Total Station.

In the event that the supervising authority employs traditional design techniques, normal practice is to supply the site engineer with the radius and the appropriate transition curve table to be used. The site engineer may also be supplied with the deflection angles and chainages in the form of a computer-generated list or file.

Setting-out transition curves

In the absence of coordinated centre-line data, the procedure for setting-out transition curves is similar to that for setting-out horizontal circular curves using deflection angles. It will be necessary to reposition the instrument for long curves. Using the table of bearings and chainages, provided or calculated, the procedure is as follows:

- set up the instrument on the origin of the spiral (TTP)
- align the instrument on a known station and set the correct bearing to that station
- turn to bearing of the tangent direction
- define the tangent's direction by pegs set at suitable distance
- set bearing of first point on curve and set peg at corresponding chainage
- continue setting-out further points until table indicates need to move instrument
- move instrument to last point set-out
- set calculated bearing of origin (TTP) and align on TTP
- continue setting bearings of further points on curve.

Note: Bearings may be given as relative or whole circle bearings, but the principles are the same. If deflection angles are given rather than bearings, these are set from the tangent at the origin.

The procedure will be simpler and quicker if a Total Station is used with coordinates calculated from the curve information.

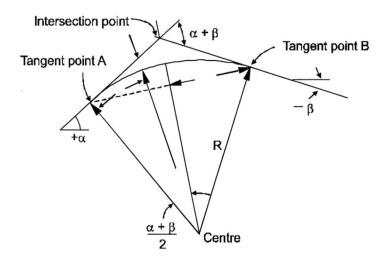

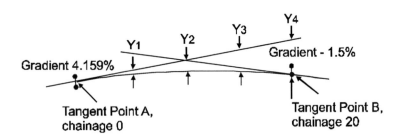

Roads: vertical circular curves

In the unlikely event that vertical curve information is not incorporated into a computer generated 3D coordinate system for setting out by Total Station, the site engineer can resort to the following processes by reference to the diagram opposite.

General equations and method

- gradients (%) to be connected ($+\alpha$, $-\beta$): convention positive (+) indicates reduced level of road increases with chainage and vice versa
- distance between tangent points, L (or radius R)
- reduced level (and coordinates) of tangent point or intersection point from which curve is to be set out.

Change of gradient $= \alpha - (-\beta) = \alpha + \beta$

For small angles, $\sin \approx \tan = \text{gradient } (\%)/100$

Hence $\dfrac{\alpha + \beta}{2 \times 100} = \dfrac{L/2}{R}$

and $R = \dfrac{100L}{\alpha + \beta}$ or $L = \dfrac{R(\alpha + \beta)}{100}$

also $y = \dfrac{x^2}{2R}$ { from $x^2 = (R + y)^2 - R^2$ neglecting y^2 }

To derive the heights of points on the curve

- select convenient intervals for x
- calculate levels for intervals along original gradient
- calculate corresponding values of y (*note:* y increases as square of interval ie in ratio 1:4:9:16:25 etc)
- deduct y to obtain height at distance x from A.

Note: If height of intersection point is given, calculate heights along original gradient as shown opposite.

Worked example

Table 2:   *Table of chainage and heights from calculated values*

CH. (m)	Point	Original grade: level change	Original grade: reduced level	y	Reduced level on centre-line
0	A	0.000	94.301	0.000	94.301
5		0.208	94.509	0.035	94.474
10	IP	0.416	94.717	0.141	94.576
15		0.624	94.925	0.318	94.607
20	B	0.832	95.133	0.566	94.567

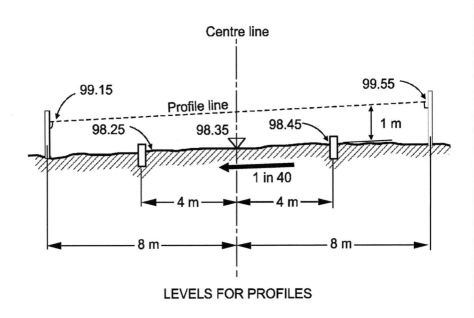

LEVELS FOR PROFILES

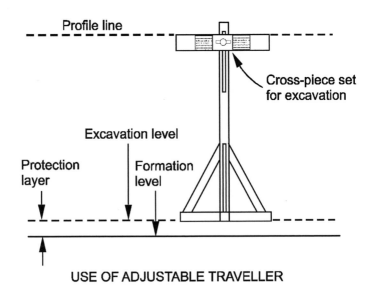

USE OF ADJUSTABLE TRAVELLER

Roads: levels

The site engineer should begin by establishing TBMs along the route, ensuring to close out the levelling run and prove the accuracy of the TBMs, as described in section Benchmarks, and then proceed as follows.

Cross sections

Set out profiles generally at 20 m intervals or preferably at 10 m intervals.

Unless the level information is provided, the site engineer will have to calculate the levels using whatever information is provided on the site drawings. All calculations should be shown on a Site Information Sheet and filed for reference.

Note: It is usually possible to position centre-line and offset pegs to coincide with the level points shown at cross- and longitudinal sections.

Profiles

For each section where profile is required:

■ determine channel levels from drawings or by calculation
■ decide offset of profiles from centre-line
■ calculate required profile levels, taking into account offset distances and adding 1 m for height of traveller (see diagram)
■ set up profiles
■ mark reduced level and chainage on profile
■ provide details of profiles on Site Information Sheets
■ check profiles daily for disturbance by sighting over three or more profiles.

Where profiles are required on existing carriageway areas (ie hard areas), or where pegs would cause an obstruction, 'stand up' profiles (similar to a traveller) can be used, set up for a particular location. These can be placed when required, then removed to allow traffic flow afterwards.

Travellers

A convenient form of traveller is shown in the diagram opposite. The height of the cross-piece can be adjusted from say, 1.5 m to 1 m as the construction proceeds.

Note: Excavation may be higher than the formation level if a protection layer is required.

After earthworks complete

After earthworks are complete, the need for accuracy in setting-out becomes imperative because of the cost of materials and final riding quality of the road's surface. Set pins or kerbs to level for sub-base, base and surface construction. Alternatively, wire lines may be required to guide paving equipment. Check and agree levels with the surfacing sub-contractor.

Suggested colour code for profiles

A suggested colour code for profiles (and pegs) is shown in the end-papers.

Section A-A

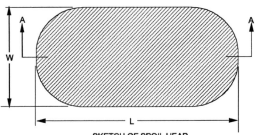

SKETCH OF SPOIL HEAP

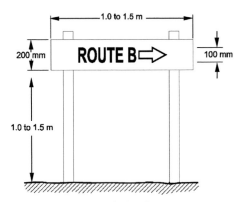

1.0 to 1.5 m

200 mm   ROUTE B ⇨   100 mm

1.0 to 1.5 m

TYPICAL DIRECTION BOARD

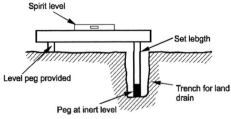

Spirit level

Set lebgth

Level peg provided

Peg at inert level

Trench for land drain

SETTING FORMATION DRAIN TO LEVEL

Roads: earthworks and drainage

Existing ground levels

Before site topsoil stripping starts:

- set up profiles at predetermined intervals
- record existing ground levels at each cross-section
- check recorded levels against those on drawings
- agree recorded levels with the supervising authority.

Note: Recording existing ground levels can be combined with setting-up profiles.

Spoil heaps

The site engineer should:

- set out the plan position of the spoil heap site with 2 m stakes painted white
- provide plan and profile of each spoil heap on an information sheet for the relevant foreman/ganger.

Haulage routes

The site engineer should:

- mark all haulage routes with direction boards that can be easily seen by the plant drivers, taking into account which public roads may not be used
- set lines and levels for routes that have to be constructed with fill to take heavy traffic.

Shoulder drainage

To control the levels of ditches and drains in shoulders:

- set up profiles as for sewers
- provide profile and traveller details on a Site Information Sheet for foreman/ganger.

Formation drains

Formation drains may be required under the road. The accuracy needed in positioning these drains is usually not great, and most of the setting-out can be left to the foreman/ganger.

The site engineer should:

- provide foreman/ganger with drain layout on Site Information Sheet
- establish pegs at suitable heights above drain levels
- make spot checks on line and level before backfilling.

Note: The ganger can transfer levels from peg to bottom of trench by means of levelling board and spirit level.

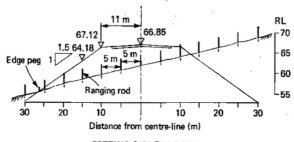

SETTING OUT EDGE PEG

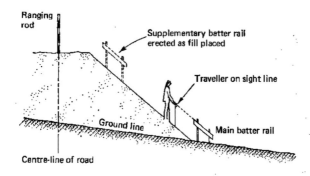

CONTROL OF SLOPE FROM BATTER RAIL

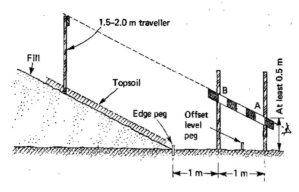

DETAIL OF BATTER RAIL

Roads: embankments

Setting-out edge pegs

Information is generally provided to the site engineer in computerised coordinate format for direct setting out of edge pegs with a Total Station. This avoids manual calculations on site, but engineers should nevertheless continue to make regular checks and not rely blindly on printout data.

Before setting-out edge pegs, the site engineer should prepare a table (as below) for each relevant chainage as in the example opposite. Calculate and enter the italicised slope line levels.

Distance from centre-line (m)	Ground level	Slope line level	Levels on batter stakes	Point
0	61.93			
5	60.97			
10	60.34			
15	59.28	64.18		
20	58.56	60.85		
25	57.21	57.52		
30	56.63	54.18		
25.5	57.15	57.18		Edge
26	57.09	56.85		
26.5	57.04	56.52	58.02	B
27.5	56.92	55.85	57.35	A

Distance up slope $= (25.5 - 11) \times \dfrac{\sqrt{(1.52^2 + 1^2)}}{1.5} = 17.43$

Setting up batter rails

At relevant chainage on site:

- set ranging rods at 5 m intervals offset from centre-line peg
- measure and record ground levels at these intervals
- by inspection and/or calculation, find distance from centreline where slope line intersects the ground line, ie the embankment edge
- mark the edge of the embankment with a peg
- set offset level peg at 1.5 m from edge of embankment
- drive two stakes 1 and 2 m from edge peg, while checking each is vertical. A taped distance may be required from point B to the top of slope line
- calculate and mark levels of point A and B for suitable traveller length (eg 1.5 m) and fix batter rail
- if the embankment is more than 5 m high, set up batter rails on the embankment slope at height intervals of 5 m as fill proceeds
- record chainage and slope distance on batter rail
- set up ranging rods on the centre-line to guide plant
- provide details of batter rails and traveller length on a Site Information Sheet.

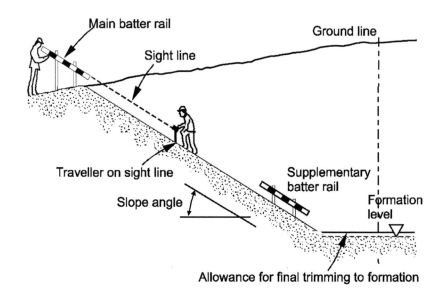

CONTROL OF SLOPE FROM BATTER RAIL

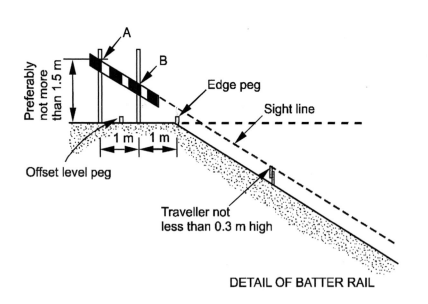

DETAIL OF BATTER RAIL

Setting-out edge pegs

To set out embankments, the site engineer should follow a similar procedure as for embankments, calculating levels on the slope line at intervals from the centre-line. The point on the embankment where the slope line intersects the ground line is marked with an edge peg.

Setting up batter rails

- set offset level peg at ca. 1.5 m for edge of cutting
- drive two profile stakes 1 m and 2 m from edge peg, while checking each is vertical. Referring to the diagram opposite, a taped distance may be required from point *B* to the top of slope line
- calculate and mark levels of points *A* and *B* for a suitable traveller length and fix batter rail
- if cutting is more than 5 m deep, set up batter rails on the slope of the cutting at depth intervals of 5 m as the cut proceeds
- record chainage and distance down slope on the batter rail
- set up ranging rods on the centre-line to guide plant
- provide details of batter rails and traveller lengths on a Site Information Sheet.

Controlling the work (cuttings and embankments)

The site engineer should:

- check batter boards, profiles and travellers regularly
- reset ranging rods on centre-line as cut or fill proceeds
- regularly check width of cut or fill either side of centre-line
- when a cut is an agreed amount above formation level (typically 300 mm) set up standard road works profiles to control final trimming and road construction
- when an embankment is nearly complete, set temporary profiles (allowing for settlement and trimming to formation level)
- when agreed by supervising authority, set final profiles for trimming to formation and pavement construction
- check the final formation level.

Note: In deep cuttings, the ground may heave due to removal of overburden. It may be necessary to check that heave has ceased before setting up the road works profiles.

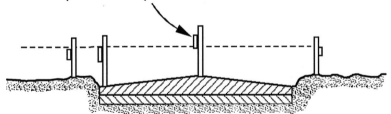

Depth of rail equals crossfall dimension

USING TRAVELLER RAIL TO CONTROL CROSSFALLS

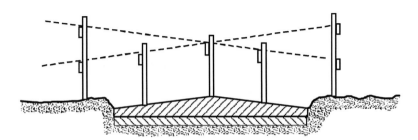

Section A: Equal crossfalls

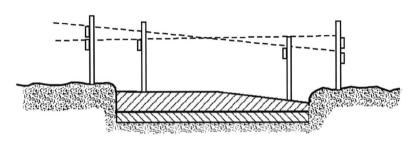

Section B: Developing super-elevation

Roads: crossfalls

It is important to shape the formation accurately as the crossfalls of the finished road are largely set by the cross-section of the formation.

Most single carriageway roads are built with crossfalls running from the centre-line so that water runs off to both sides of the carriageway. In the case of a dual carriageway, these are usually constructed with a single crossfall.

Setting profiles

In order to sight across the road, the site engineer sets profile boards parallel rather than perpendicular to the centre-lines. Profiles should be set at say, 10 m intervals.

Using traveller rail to control crossfalls

For narrow or minor roads, it may be sufficient to check the levels at the centre-line and road edges only and to check in between these levels by eye or with a straight-edge.

- make the depth of the traveller rail equal to the crossfall
- sight to the top of the traveller rail at the road edges
- sight to the underside of the traveller rail at the centre-line.

Using profiles rail(s) to control crossfalls

More usually, it is necessary for the site engineer to check levels right across the section. In general, set two rails on each profile, so that the sight line from the upper rail on one side to the lower rail on the far side, controls the far crossfall and vice versa (*see* Section A of diagram).

When developing super-elevation, the two profile rails are positioned closer (*see* Section B of diagram), and it may be necessary to use two profiles side by side.

When the super-elevation is fully developed (*see* Section C of diagram), this provides the crossfall and each profile needs only a single profile rail.

Supervision

The site engineer should:

- give details of the profiles and travellers on a Site Information Sheet to the appropriate foreman/ganger
- check that the profiles and traveller are used correctly.

Using road pins to control surfacing

Line and level are usually controlled using road pins once the construction is sufficiently advanced. The pins are set at regular intervals (say 10 m on straights – closer to control tight curves) either on the kerb line or at a suitable offset from the carriageway edge. If the road has no kerb, the offset must allow sufficient width to allow the surfacing plant access and for any stepping-out of the lower layers.

Normal practice is to mark level values on the pins using tape. Crossfalls can be shown using different colour tapes. Levels across the carriageway are checked by stretching string lines between pairs of pins.

It is common practice for the levels given on the pins to be at a predetermined height above finished road level. This is termed the 'block up' and may be set to kerb height or to any other agreed figure (eg 100 mm).

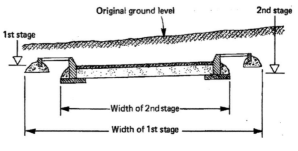

TWO-STAGE EXCAVATION

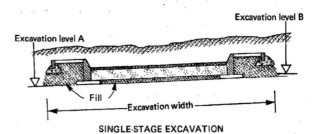

SINGLE-STAGE EXCAVATION

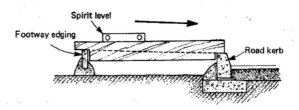

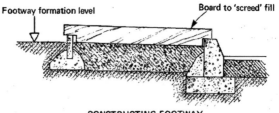

CONSTRUCTING FOOTWAY

Roads: footways and kerbs

When roads incorporate footways, excavation can be undertaken in two stages or a single stage. Site management will make the choice, depending on such factors as plant to be used, and costs of removing excavated material and importing fill material.

Two-stage excavation

- set up profiles 1 m beyond each footway
- excavate over full width of road and footways to footway formation level
- reset profiles 1 m beyond each kerb line
- excavate over road width only to road formation level (boning along and across road to control crossfall)
- excavate as necessary along kerb line for kerb bed
- excavate for footway edging.

Note: This method is best-suited to short lengths of road where heavy earthmoving plant is not appropriate.

Single-stage excavation

- set up profiles 1 m beyond each footway
- excavate over full width of road and footway to kerb foundation level (level A) or road formation level (level B) as preferred (the former choice may require importing additional fill to road formation level).

Note: This method is suitable where heavy earthmoving plant is justified by scale of the job.

Kerb and channel levels

The site engineer should:

- establish pins at kerb level on the required line (front of kerb, back of kerb or agreed offset). These are usually set at 10 m intervals on straights or moderate curves; 5 m on tight curves
- if channel required, because there is no longitudinal fall, mark each 'valley' and 'summit' on face of kerb in waterproof chalk. Kerb layer will use chalk line between valley and summit to lay concrete channel
- provide foreman/ganger with details of any offsets or block ups used on a Site Information Sheet.

Constructing footway

Crossfall and width of footways can be controlled by using notched board and spirit level as shown. Fill between kerb and edging as necessary and 'screed' to formation level.

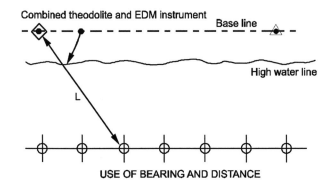

USE OF BEARING AND DISTANCE

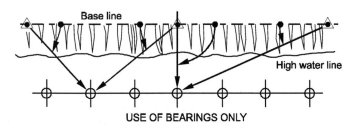

USE OF BEARINGS ONLY

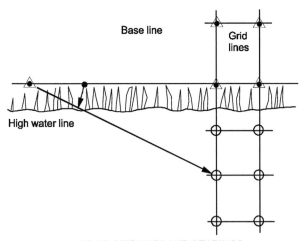

USE OF GRIDLINES AND BEARINGS

River works: bearing piles

River and marine works often as not require specialist knowledge particularly regarding checking soil condition and working on floating plant and unstable ground.

In this short work, only the installation of piling is considered.

Bearing piles are taken to include piles capable of resisting lateral loads (eg mooring loads) as well as vertical loads. Such piles may be driven as raking piles.

Method of installation

It is usual to drive temporary piles from a moored floating platform and then to erect a temporary staging from which to pitch and drive the permanent piles. During driving, each pile is held in a 'gate' (or guide) to achieve the specified tolerances for location and rake; of necessity these tolerances are commonly less onerous than for land-based piling.

The difficulties

The major difficulties in setting-out are measuring distances over water and providing stable reference points. A (temporary) reference pile can incorporate a sighting mark but, if so, it is important to check for oscillation of the pile, particularly in fast currents.

Check that sight lines to or from water level will be unobstructed whatever the state of the tide or water level.

Communication can be difficult and a radio link is advisable.

Locating piles by bearing and distance

Probably the most convenient method of locating piles over water is to use a Total Station of GPS system.

Locating piles by bearing only

For lines of piles roughly parallel to a river bank, an alternative method may be to use bearings only from an appropriate pair of reference points. The disadvantage of this method is that 'three-way' communication is required.

Locating piles with grid lines and bearings

For piles in lines roughly perpendicular to a river bank, where practicable:

- establish reference points defining these grid lines
- set up reference point some distance along the bank and calculate bearing of each pile from this point
- for each pile use two theodolites/Total Stations to set out the grid line and the intersecting bearing.

Note: If piles are constructed from the river bank, tape along grid line on the temporary staging used for piling.

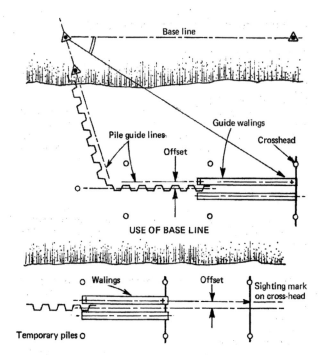

USE OF BASE LINE

USE OF SIGHTING MARK

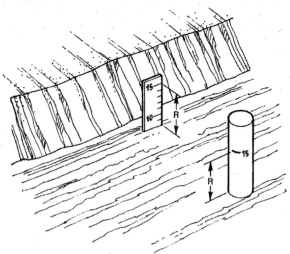

USE OF STAFF GAUGE

River works: sheet piles

Method of installation

As on land, sheet piles are driven in 'panels', using a pair of walings as a guide. Pairs of temporary piles are driven at 'panel' intervals either side of the sheet pile centre-line. A crosshead, located on a pair of temporary piles, supports the 'leading' end of the walings. The 'trailing' end is supported on the end pile of the panel of sheet piles previously driven. Thus it is necessary to check the line of the leading end of the waling before each panel is driven.

Note: Temporary piles are often later used to support falsework.

Line of piles at angle to shore

For sheet piles at an angle to a river bank:

- set up theodolite or total station on line offset to centre-line of piles; fix offset targets at each end of waling
- sight along offset centre-line and align both ends of waling
- after driving first panel of piles, set trailing end of waling against last pile driven and align leading end.

Line of piles roughly parallel to river bank

Set up a baseline along the river bank and use bearing and distance or two bearings to set up the walings as described previously. Alternatively, sighting marks could be set up on temporary piles.

Driving piles to level

To drive piles to required level with fair accuracy in reasonably calm water:

- set up staff gauge to be read from piling rig (staff gauge indicates tide/river level relative to the level datum)
- mark required level on pile
- drive pile until mark on pile and same level on staff gauge are same distance above water level.

Underwater concreting

Having constructed a sheet pile cofferdam, it may be necessary to concrete a base underwater. If formwork is required, it must be positioned by divers. Within reasonable limits of accuracy, this can be achieved as follows:

- set out each locating point from the shore stations
- fix plywood sheets (or timbers) to the upper walings
- drill small holes through ply and suspend plumb-bobs from piano wires passed through the holes
- when divers have located formwork by plumb-bobs, use weighted tape to define required concreting level.

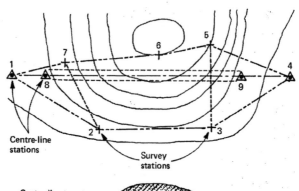

Centre-line
stations

Survey
stations

Centre-line target

TUNNEL WITH SHALLOW GRADIENT

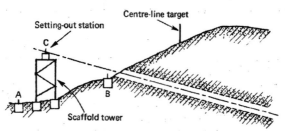

Centre-line target

Setting-out station

C

A

B

Scaffold tower

TUNNEL WITH STEEP GRADIENT

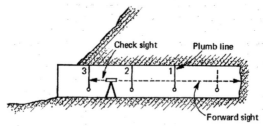

Check sight   Plumb line

3   2   1

Forward sight

PROJECTING CENTRE-LINE FORWARD

Tunnelling: from ground level

Tunnelling is a specialised activity and demands particular attention to detail. The following therefore applies to minor works and may be inappropriate for major tunnelling projects.

Tunnel with shallow gradient

From the diagram, assuming that points 1 and 4 define the centre-line:

- determine the bearing of 4 from 1 by carrying out a closed traverse
- set up theodolite/Total Station on station 1, set bearing of survey station 2 and align telescope on that station
- set out centre-line station 8 for portal construction
- set out centre-line target above and clear of future tunnel works to replace station 8 when portal complete
- when section of tunnel constructed, transfer centre-line on to roof at three well-spaced sections.

Tunnel with steep gradient

Assuming that centre-line stations A and B have been set out:

- set up centre-line target above tunnel works
- erect heavily braced scaffold tower between A and B
- set out station C on top of scaffold tower
- with theodolite/Total Station on station C, align on centre-line target, and set vertical angle to gradient of tunnel to provide line parallel but offset vertically to centre-line
- check vertical offset by levelling at tunnel portal.

Note: Line and gradient can subsequently be controlled by laser but check frequently with theodolite or Total Station.

Projecting centre-line forward:

- suspend plumb lines from three centre-line marks nearest tunnel face (*see* illustration)
- set up theodolite between plumb lines 2 and 3
- sight towards face and traverse telescope until offsets of plumb lines 1 and 2 are equal
- traverse through 180° to check offset of plumb line 3
- if offsets agree, traverse again through 180° and set one or more new (offset) centre-marks on tunnel roof
- if offsets disagree, re-check suspect centre-marks from proven centre-marks further from face.

Note: Lasers are now commonly used to control lines. A single laser may suffice for tunnels driven by tunnel boring machines, but for other methods three may be used, one in the crown and one either side just above centre at suitable offsets, to control both line and tunnel profile.

Need for accuracy

Use a theodolite or Total Station reading to 1 second and average a number of readings.

Note: An error of 20 seconds is equivalent to an alignment error of about 100 mm per kilometre.

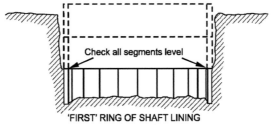

Check all segments level

'FIRST' RING OF SHAFT LINING

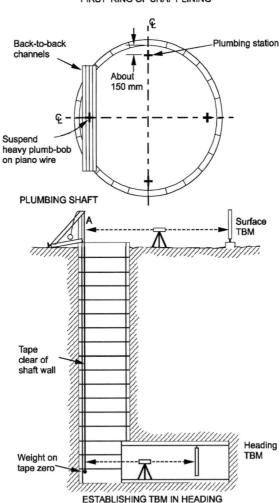

C̵L

Back-to-back channels

Plumbing station

About 150 mm

C̵L

C̵L

Suspend heavy plumb-bob on piano wire

PLUMBING SHAFT

A

Surface TBM

Tape clear of shaft wall

Heading TBM

Weight on tape zero

ESTABLISHING TBM IN HEADING

Tunnelling: constructing shafts

Setting-out first ring

Most shafts are circular and lined with pre-cast concrete or cast iron bolted segments. The first ring constructed is usually the third ring down.

When 'first' ring has been set up with bolts hand-tight only:

- re-establish centre-point
- check concentricity of the segments
- check uniformity of level of all segments.

    The bolts can then be tightened, the rings above constructed and a concrete 'collar' placed for stability. If these first rings are set and maintained true, subsequent adjustments for line and level need only be minimal.

Plumbing down shaft

Check verticality of shaft as follows:

- establish four plumbing stations about 150 mm from shaft wall at ends of two orthogonal centre-lines
- suspend plumb lines from these stations (damp movement of bobs by suspending in oil or water)
- in deep shafts, where visibility is poor, check that plumb line is clear of shaft well by dropping snap washer down plumb line
- check verticality by measuring offset at 'first' ring and last ring constructed.

Alternatively, a suitable optical plumbing instrument or laser may be set up over each plumbing station.

Establishing TBM in heading

- suspend weighted tape from timber frame at top of shaft
- set up level at surface, sight first on surface TBM, determine collimation level and read tape at A
- set up level in heading, read tape at B, and calculate collimation level
- determine reduced level of TBM on roof of heading.

Note: Weight on tape should be calibration load less half the self weight of the suspended portion of the tape. Where possible, use full depth tape. If this is not practicable, fix suspension points at suitable intervals down shaft and tape between these. For very deep shafts, electronic/laser distance measurement equipment may be appropriate.

Warning: Shot-fired pins are convenient for TBM and other levelling points in rock tunnels but check safety aspects, especially methane/gas risks.

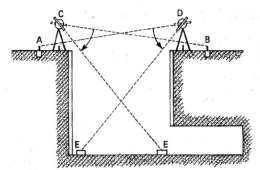

TRANSFERRING LINE DOWN SHALLOW SHAFT

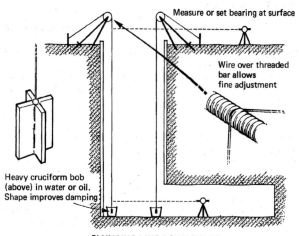

Measure or set bearing at surface

Wire over threaded bar allows fine adjustment

Heavy cruciform bob (above) in water or oil. Shape improves damping

PLUMBING LINE DOWN SHAFT

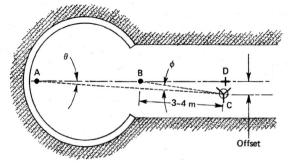

3–4 m

Offset

WEISBACH TRIANGLE

Transferring line down shallow shaft

With A and B defining the required heading direction:

- set up subsidiary stations C and D close to shaft
- from C and D, set out F and E respectively.

Note: Check accuracy by repeating observations.

Plumbing line down shaft

Two plumb lines are suspended down the shaft, as far apart as practicable. The two lines may define:

- given bearing or direction, or
- random bearing which is measured at the surface.

'Picking up' bearing

Whether at the surface or underground, the bearing defined by the plumb lines has to be 'picked up' by using a 1-second theodolite/Total Station and the Weisbach triangle (*see* below). Underground, illuminate each plumb line (piano wire) with a lamp. A scale fixed behind the line will help in eliminating any residual movement by observing the mean position.

Use of the Weisbach triangle:

- set up instrument at C, about 3 to 4 m from B and offset by say 25–50 mm from the sine defined by AB
- measure the angle subtended by A and B, φ (average of a number of readings on both faces)
- measure the lengths AB and BC
- calculate $\theta = \phi \; \dfrac{BC}{AB}$ where θ and ϕ are in seconds
- calculate offset $CD = \dfrac{\theta \, (AB + BC)}{57.2958 \times 3600}$
- align on A and traverse through 180°− θ (clockwise as illustrated) to sight along line by calculated offset
- mark centre-line by taping from offset line (appropriate for small tunnels or tunnels with bends) or
- mark further line offset from centre-line by convenient amount up to, say, 1 m (appropriate for large tunnel where centre-line may be obscured by ventilating duct).

Gyro-theodolite

A gyro-theodolite can provide an accurate azimuth where other methods would be impossible. It should be noted that gyro-instruments track True North and not grid or magnetic north, hence a significant angular correction will be necessary. Setting up and operating a gyro-theodolite is a specialist operation and must not be undertaken by anyone who is not familiar with the concepts or competent to make the necessary adjustments.

Controlling squareness

In large tunnels where a primary lining is used, the control of line is related to the control of the 'squareness' of the lining. To check squareness:

- set up the instrument on (offset) centre-line
- turn through 90° (left and right)
- rotate telescope to 45° above and below horizontal and mark lining
- tape from the four marks to lining at face.

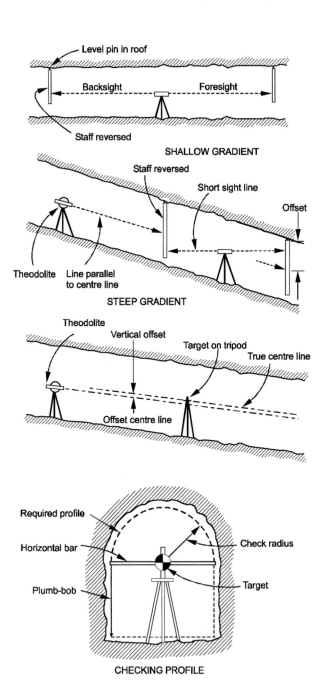

Level pin in roof

Backsight   Foresight

Staff reversed

SHALLOW GRADIENT

Staff reversed

Short sight line

Offset

Theodolite   Line parallel to centre line

STEEP GRADIENT

Theodolite

Vertical offset

Target on tripod

True centre line

Offset centre line

Required profile

Horizontal bar

Check radius

Plumb-bob

Target

CHECKING PROFILE

Controlling gradients

Gradients may be controlled by:

- theodolite/Total Station set to required gradient
- profiles fixed to roof (moderately convenient for excavation by drilling and blasting or hand digging)
- laser aligned to gradient, established by conventional methods and convenient for all methods of excavation now commonly used.

Checking levels on shallow gradient

Check gradient and reduced levels at regular intervals:

- establish underground TBM and level points in roof
- level from underground TBM using inverted staff.

Note: TBM and level pins may have to be offset from centre-line to avoid ventilation ducts, etc.

Checking levels on steep gradient

Standard tunnel levelling procedure may be used but note that short sight lines tend to reduce precision and any errors will accumulate rapidly, eg if back-sight and fore-sight distances are not kept equal. Alternatively, check with theodolite/Total Station:

- set vertical circle on instrument to gradient
- using inverted staff, check offset of known roof level
- compare offsets of pins further forward.

Profile of cross-section

The site engineer may have to check selected cross-sections for measurement purposes or because ground movement is suspected. A suitable technique is:

- align instrument on offset centre-line and to gradient (for convenience, directly below a proven level)
- measure vertical offset from centre-line
- set up tripod and target at selected cross-section
- align target with offset centre-line
- adjust target height by offset distance to establish centre-point of section
- check profile dimensions by taping or using a trammel.

Note: A horizontal cross bar may be necessary if the cross-section is non-circular (*see* illustration).

Where the tunnel is controlled by lasers, the profile may be checked by direct measurement of offsets from the laser beam, hold the tape on the beam and move it back and forth to obtain the smallest reading on the tape.

Controlling length

Controlling length may be critical if bends or other features must be located precisely. An invar band, rather than a steel tape, may be necessary together with catenary taping techniques, using firm supports with well-defined datum. When using any form of electronic distance measurement in a tunnel, the greatest care needs to be taken to avoid introducing temperature and refraction errors (refraction will cause the light beam to bend and hence give a distance over great).

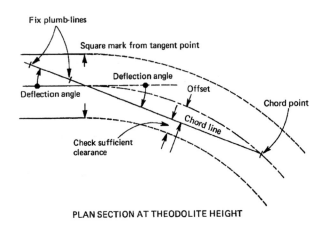

PLAN SECTION AT THEODOLITE HEIGHT

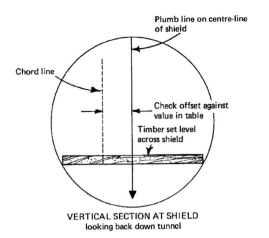

VERTICAL SECTION AT SHIELD
looking back down tunnel

Tunnelling: around curves

Office procedure

- on plan of tunnel, determine suitable chord(s) that will not be too close to tunnel wall along line of sight and calculate deflection angle
- on plan, mark off ring width on inside radius of tunnel
- calculate offset from chord to centre of each ring
- calculate required lead per ring on outside radius (lead arises from greater arc length on outside of curves).

Procedure in the tunnel

- with theodolite/Total Station at tangent point, sight back down tunnel on centre-mark and traverse through deflection angle (alternatively, set out using coordinates)
- set up two plumb lines on chord line in straight portion of tunnel
- set out square marks from tangent point
- as excavation proceeds extend chord line into curved portion and set up plumb lines
- check offsets from chord to centre of ring/shield
- when shield is clear of first chord point, set out this point accurately by bearing and distance
- set up on first chord point, sight on tangent point and traverse to set up second chord line
- set out square marks at first chord point
- check any deviation of the tunnel at the first chord point and distribute the error to bring tunnel back onto true centre by or before the second chord point
- repeat establishment of new chords as necessary.

Site Information Sheet

A Site Information Sheet for the foreman/ganger should be prepared by the site engineer and include a table with headings as below:

Ring No.	Offset (mm)	Lead from mark	
		Ring No. marked	Distance (m)

Tunnels are now commonly controlled by laser; similar principles apply.

Appendix A: Contractural aspects

Responsibilities

The responsibilities of the supervising authority in respect of setting-out the Works, are normally laid down in the Conditions of Contract. Many types of contract are currently in use in the construction industry, but the roles and responsibilities set forth in the conditions reflect the form of procurement used by the client.

It is important to establish who bears responsibility for providing information, setting-out the works, rectifying errors and checking.

The following points are typical of some forms of contract (ICE Conditions or JCT Standard Form):

- the contractor bears the responsibility for setting-out the works
- the contractor generally bears the cost of rectifying errors arising from incorrect setting-out
- the contractor must not rely on any checking of the setting-out by the engineer or the architect or their representatives
- the engineer, architect or supervising officer must provide, in writing, essential data required by the contractor for setting-out the works
- the contractor may engage a specialist firm to carry out the main setting-out operations but should check the results independently
- similarly, the contractor may agree that a sub-contractor shall set-out his part of the works but should check the results independently and at frequent intervals.

Typical procurement routes

The choice of procurement system is widely varied. A key defining feature is the way in which responsibility is allocated by the client.

One can distinguish between:

- multi-point systems, in which a number of organisations are separately responsible for design and construction
- single point systems, in which a single organisation takes complete or substantial responsibility for both design and construction.

The following is abstracted from section 6 of CIRIA publication SP113 *Planning to build? – A practical introduction to the construction process*.

Traditional designer-led procurement

This is a multi-point approach with the design team and the contractor both separately appointed by the client. There is no direct contractual link between the design team and the contractor.

Management contracting

A similar contractual arrangement to the traditional approach, except that the contractor is paid on a fee basis to manage a series of sub-contract packages to carry out the works.

Construction management

A procurement system in which a construction manager coordinates and directs 'trade contractors', each of which has a contract directly with the client.

Design and build

A single point system, whereby the client appoints a design and build organisation to manage both the design team and the construction team. These teams may be a mixture of in-house capability and sub-contractors.

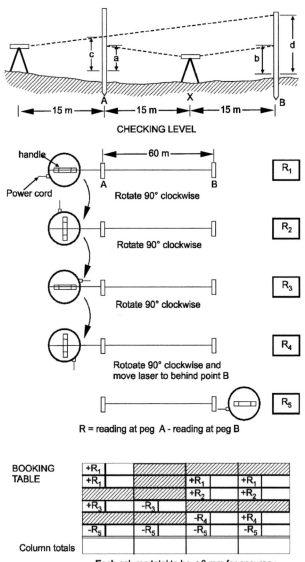

CHECKING LEVEL

handle

Power cord

A — 60 m — B

R_1

Rotate 90° clockwise

R_2

Rotate 90° clockwise

R_3

Rotate 90° clockwise

R_4

Rotoate 90° clockwise and move laser to behind point B

R_5

R = reading at peg A - reading at peg B

BOOKING TABLE

$+R_1$							
$+R_1$				$+R_1$		$+R_1$	
				$+R_2$		$+R_2$	
$+R_3$		$-R_3$					
				$-R_4$		$+R_4$	
$-R_5$		$-R_5$		$-R_5$		$-R_5$	

Column totals

Each column total to be < 6 mm for accuracy

CHECK ROTATING LASER

Care and maintenance

Site engineers are responsible for the day-to-day care and maintenance of instruments. All setting-out instruments should be packed and stored in secure, dry conditions when not in use. After use, if wet, instruments should be wiped down and left to air-dry before packing away. Lenses should be regularly cleaned using lens tissue or a dusting brush. Instruments (and tripods) are expensive and must not be left on site unattended, even for short periods. Damage or total destruction is often caused by mobile plant. When carrying instruments by road, ensure they are secured in the vehicle. Avoid carrying instruments out of their boxes over long distances and avoid carrying them still attached to the tripod as this places unnecessary strain on the bearings.

Adjusting instruments

When adjusting instruments, always follow the manufacturers' instructions.

Checking level

A level should be checked at least weekly (two-peg test):

- from X, take readings a and b
- apparent difference in level of A from B = a − b
- from Y, take readings c and d
- if [(a − b)−(c − d)] is more than say, 3 mm, the instrument requires adjustment.

Checking theodolite or Total Station

Plate bubble: Set up the instrument until the bubble tube is parallel with two foot screws, and level the bubble. Turn instrument through 90°, level bubble using third foot screw. Turn back through 90°, re-level bubble if necessary. Then turn through 180°, the bubble should still be central if correctly adjusted. If necessary, adjust according to manufacturer's instructions.

Optical plummet – two types:

1 Fixed in tribrach:
  - level tribrach with instrument or plate level over point defined by a cross. With a sharp pencil, trace the outline of the tribrach base plate on the tripod head. Carefully turn the tribrach through 120° keeping within the marks on the tripod. Re-level and sight the cross. Repeat, turning through a further 120°. The cross should be central each time.

2 Revolving plummet fixed in instrument or plate level.
  - level and centre over a fixed cross and rotate through 120° and 240° observing the cross, which should stay central each time. Horizontal circle: set up and level the instrument. Sighting a well-defined distant point, read the horizontal angle scale (H1). Transit the telescope and re-sight the point on the opposite face, read the horizontal angle scale (H2). The difference between H1 and H2 should be 180°.

Vertical circle:

  - set up and level the instrument. Sighting a well-defined elevated point, read the vertical angle scale (V1). Transit the telescope and re-sight the point on the opposite face, read the vertical angle scale (V1). (V1) and (V2) should be equal.

Trunnion axis:

  - set up and level the instrument. Sight a well-defined high-level point with the centre of the cross-hairs. Depress the telescope and read a tape set close to the instrument (T1). Repeat this on the opposite face and read the tape again (T2). The readings (T1) and (T2) should be equal.

Checking rotating laser

Carry out 'two-peg' test as illustrated opposite and adjust according to the manufacturer's instructions.

Local scale factor for any part of England, Scotland or Wales

National Grid Easting (km)		Scale Factor F
400	400	0.99960
390	410	0.99960
380	420	0.99961
370	430	0.99961
360	440	0.99962
350	450	0.99963
340	460	0.99965
330	470	0.99966
320	480	0.99968
310	490	0.99970
300	500	0.99972
290	510	0.99975
280	520	0.99978
270	530	0.99981
260	540	0.99984
250	550	0.99988
240	560	0.99992
230	570	0.99996
220	580	1.00000
210	590	1.00004
200	600	1.00009
190	610	1.00014
180	620	1.00020
170	630	1.00025
160	640	1.00031
150	650	1.00037
140	660	1.00043
130	670	1.00050

Appendix C:
National Grid, benchmarks and ground distances

National grid

All Ordnance Survey Geographical Information is based on the National Grid of Great Britain. Historically, National Grid was defined by physical triangulation stations connected to the primary triangulation of Great Britain. National Grid is now established from GPS-derived coordinates plus a transformation model. The GPS coordinates are defined in European Terrestrial Reference System 1989 (ETRS89). ETRS89 is a well-defined and stable coordinate system fully compatible with WGS84 (the coordinate system used by GPS), and has been adopted as the primary system for accurate coordinate positioning across Europe.

A GPS receiver on a control point on site is connected, either in real-time or after the fact through post processing, with permanent GPS stations that make up the Ordnance Survey's OS Net™ network, or a passive station that has previously been coordinated with GPS. By doing this, the local control is automatically in ETRS89. A transformation model (OSTN02™) is then used to convert the coordinate to National Grid. OSTN02 may be updated in the future by Ordnance Survey. An online coordinate converter is available at <http://gps.ordnancesurvey.co.uk/convert.asp>.

Data, coordinates and information on the National GPS Network and OS Net is freely available at the Ordnance Survey GPS website <www.ordnancesurvey.co.uk/oswebsite/gps>. Historical triangulation control coordinates are still available from: <http://benchmarks.ordnancesurvey.co.uk>.

Ordnance Benchmarks

Traditionally, bringing Ordnance Datum Newlyn, or other Ordnance Survey height datum, into a site was achieved through spirit levelling between local OS benchmarks and a control point on site. This has now been superseded by GPS in a similar manner to how National Grid is brought into a site.

A GPS receiver on a control point on site is connected, either in real-time or after the fact through post processing, with one of the permanent GPS stations of the Ordnance Survey's OS Net™ network or a passive station that has previously been coordinated with GPS. By doing this, the local control is automatically in ETRS89. A Geoid model (OSGM02™) is then used to convert the coordinate to a height above Newlyn, or other Ordnance Survey height datum in the Scottish islands. OSGM02 may be updated in the future by Ordnance Survey.

Once a benchmark has been established, spirit levelling can be used to transfer that level to temporary benchmarks (TBMs).

Data, coordinates and information on the National GPS Network and OS Net is freely available at the Ordnance Survey GPS website <www.ordnancesurvey.co.uk/oswebsite/gps>. Historical benchmark values are still available from: <http://benchmarks.ordnancesurvey.co.uk>.

Ground distances

The distance calculated between any two National Grid coordinates is the projection distance and differs from the ground distance (ie the true horizontal distance). To convert projection distance to ground distance, it is necessary to use a local scale factor.

Ground distance = (projection distance/local scale factor).

Example:
National Grid coordinates

Station	Eastings	Northings
B	482841.570	363542.210
A	482671.950	363433.540
Difference	169.620	108.670

Projection distance $= \sqrt{169.620^2 + 108.670^2}) = 201.445$ m

Ground distance $= \dfrac{201.445}{0.99968} = 201.510$ m

Note: This is the ground distance at mean sea level; a further correction of +15.7 ppm per 100 m of height above mean sea level is needed to give the true ground distance.

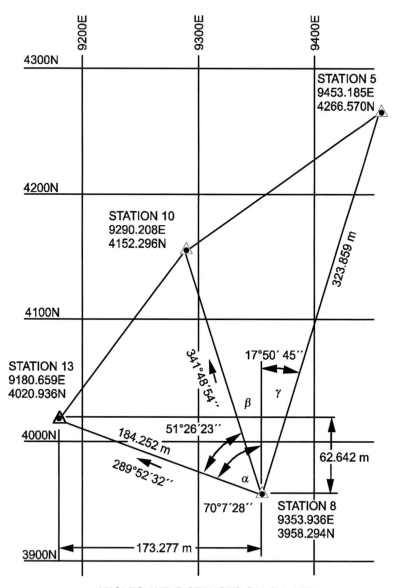

ANGLES AND DISTANCES CALCULATED
FROM GIVEN CO-ORDINATES

Appendix D:

Road layout: traditional method Proving setting-out stations

The procedure for proving the coordinates of setting-out stations is given in the main text. An example is given below using the road layout illustrated previously.

Calculate ground distance

From coordinates of stations 8 and 13, for example:

Station	Eastings	Northings
Station 13	9180.659	4020.936
Station 8	9353.936	3958.294
Partial coordinates 8 to 13	−173.277	+62.642

Calculated ground distance 8 to 13 $= \sqrt{(173.277^2 + 62.642^2)}$ = 184.252 m

Check ground distance

Assume that measured and calculated ground distances between stations 8 and 13 agree within 20 mm.

Therefore Stations 8 and 13 may define baseline.

Calculate relative bearings

Referring to figure opposite, for stations 8 and 13:

tan α = (δ eastings/ δ northings) = 173.277/62.642, α = 70° 07' 28"

Therefore, whole circle bearing = 289° 52' 32"

Repeat for Stations 10 and 8

Station	Eastings	Northings
Station 10	9290.208	4152.296
Station 8	9353.936	3958.294
Partial coordinates 8 to 10	−63.728	+194.002

tan β = 63.728/194.002, β =18° 11' 06", WCB = 341° 48' 54"

Repeat for Stations 5 and 8, γ = whole circle bearing = 17° 50' 45"

Note: The calculation of ground distance and bearing is facilitated by the use of polar coordinates on a suitable calculator or in a spreadsheet but the above method would provide a suitable cross-check.

Check relative bearings

- set up instrument on station 8
- set upper plate to whole circle bearing for station 13 from station 8
- align instrument on station 13 and lock lower plate
- release upper plate and observe whole circle bearings when aligned on stations 10 and 5 in turn.

Repeat procedure on opposite face. Check that the error between mean observed and calculated whole circle bearings is acceptable.

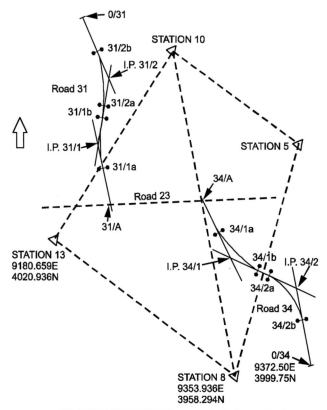

CO-ORDINATE LAYOUT (some co-ordinates omitted)

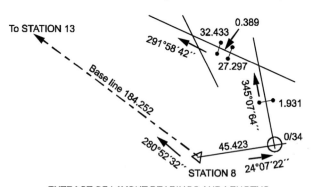

EXTRACT OF LAYOUT BEARINGS AND LENGTHS

Appendix D:
Road layout: traditional method for setting-out main points

When the setting-out stations have been 'proved' on site, the site engineer can calculate the bearings and distances of the main points and centre-lines of roads 34 and 31 in the layout. This is facilitated by drawing skeleton layouts as shown opposite. The top layout shows the centre-lines and (some) coordinates of centre points and intersection points. Tangent points are 34/1a, 34/1b, etc. The lower layout is a basis for recording calculated bearings and distances for use on site.

The calculations should be done in the office.

Starting from Station 8, to fix the base point of the hammer head on road 34 (0/34), coordinates are converted into bearings and distances thus:

Station	Eastings	Northings
0/34	9372.500	3999.750
Station 8	9353.936	3958.294
Partial coordinates 8 to 0/34	+18.564	+41.456

α = whole circle bearing = $\tan^{-1}$ (18.564/41.456) = 24° 07' 22"

The distance from 8 to 0/34 = $\sqrt{(18.564^2+41.456^2)}$ = 45.423 m

Note: The calculation of ground and bearings is facilitated by the use of polar coordinates on a suitable calculator or in a spreadsheet but the above method would provide a suitable cross-check.

Similar calculations are repeated along the main line of roads 34 and 31. To ensure the setting-out 'closes', check bearings and distances from alternative stations are added.

The setting engineer can now start setting-out the main points on site using colour coded pegs as described previously. Setting up the instrument over station 8, fix bearing 289° 52' 32" and sight on station 13. Then reading bearing 24° 07' 22", measure 45.42 m, taking into account slope. This will determine point 0/34. This sequence is repeated along the main lines of roads 34 and 31 starting at 0/34. Check bearings are taken from station 13 and station 10 to 0/31.

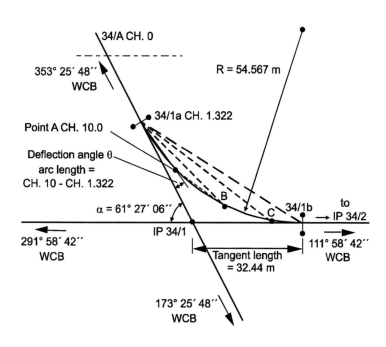

Table 3 *Setting-out centre-line of curve*

Station	Chainage of chord PT m	Arc length m	Deflection angle			WBC Reading			Chord length m
			Deg	Min	Sec	Deg	Min	Sec	
(1)	(2)	(3)		(4)			(5)		(6)
34/1a	1.322	8.678	04	33	22	173	25	48	
'A'	10.000	10.000	05	15	00	168	52	26	8.669
	20.000	10.000	05	15	00	163	37	26	9.986
	30.000	10.000	05	15	00	158	22	26	9.986
	40.000	10.000	05	15	00	153	07	26	9.986
	50.000	9.847	05	10	11	147	52	26	9.986
34/1b	59.847					142	42	15	9.834
		$\alpha/2 =$	(30	43	33)	(111	58	42)	
		$a =$	61	27	06				

Appendix D:
Road layout: traditional method: setting-out horizontal curves

This example is based on one of the curves of the estate road layout given previously for which the coordinates and bearings of the skeleton have been worked out.

Given, in addition to the tangent bearings:

$R = 54.567$ m, tangent length $= 32.44$ m

Deflection angle of point A and bearing of tangents

Arc length 34/1a to A = CH 10.00 − CH 1.322 = 8.678 m = $R \times 2\theta$ (θ in radians)

Deflection angle θ (deg) = 04° 33' 22"

Whole circle bearing of A from 34/1a

$= 173° 25' 48" − 04° 33' 22" = 168° 52' 26"$

Chord length 34/1a to $A = 2R \sin\theta = 8.669$ m

Repeat for subsequent points at 10 m chainage intervals (arc length) and for point 34/1b, as shown in the table.

Using the proforma shown opposite, calculate all the deflection angles at all the chosen points along the arc (column 4). The sum of these deflections should equal $\alpha/2$. The whole circle bearing is then worked out to each of the chainages.

Assuming instrument is set up at 34/1a, final bearing to 34/1b =142° 42' 15"

To check:

142° 42' 15" = bearing of the far tangent point from 34/1a

IP 34/1 to IP 34/2 = 291° 58' 42" − 180°= 111° 58' 42"

Final bearing   = 142° 42' 15" − sum of deflections

               = 142° 42' 15" − 30° 43' 33"

               = 111° 58' 42"

Finally calculate chord lengths (column 6)

The instrument remains at 34/1a, the chord distances are from intermediate point to intermediate point rather than calculating and taping the long chords from 34/1a to each intermediate point. On a long curve or if the sight line is obstructed it may be necessary to move the instrument to one of the intermediate points, setting it to the back bearing of point 34/1a and then proceeding as before.

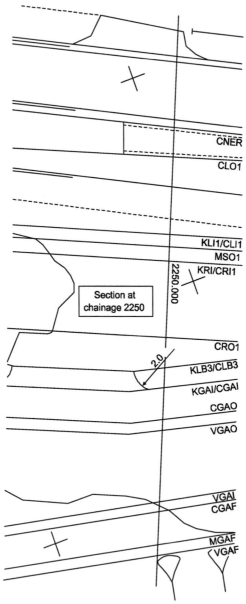

Part of computer printout derived from MOSS data

Appendix E: MOSS data for roads

Data for the construction of roads and associated structures are often provided in the form of MOSS data strings. For long sections, points on the string are given coordinates (x, y and z), chainage, whole circle bearing of the line at that point and radius of vertical curvature. Each longitudinal string is given a reference code. For cross sections at a given chainage, points are defined in terms of their coordinates (x, y and z), an offset distance, usually from the centre-line, and a reference to the codes of the longitudinal strings.

In common with all computer software based design and build packages, formats and protocols, MOSS is subject to change and improvement. Indeed, MOSS is not the only format and the setting-out engineer is encouraged to keep abreast of developments.

Dated examples of data for longitudinal and cross sections are given below and opposite. When checking the data provided, ensure that the various strings etc. have been properly identified and clearly understood.

Table 4 *Sample of main line MOSS data, as a cross-section*
Model section report: Chainage 2250

POINT	X	Y	Z	OFFSET	LABEL
					CUT
3	7582.732	7074.441	188.239	−11.568	CNER
4	7581.722	7071.597	188.241	−8.550	CLO1
5	7579.278	7064.718	188.423	−1.250	CLI1 (Channel line)
6	7579.275	7064.709	188.523	−1.240	KLI1 (Channel line, offset 1.25 m)
7	7578.860	7063.540	188.423	0.000	M501 (Top of kerb, offset 1.24 m)
8	7578.445	7062.372	188.523	1.240	KRII (Main line)
9	7574.442	7062.362	188.423	1.250	CR I1 (Top of kerb)
10	7575.998	7055.483	188.241	8.550	CRO1 (Channel)
11	7575.073	7052.880	188.172	11.312	CLB3 (Channel)
12	7575.070	7052.871	188.272	11.322	KLB3 (Channel in lay-by)
13	7574.407	7051.005	188.323	13.302	KGAI (Kerb in lay-by)
14	7574.404	7050.996	188.273	13.312	CGAI

Table 5 *Sample of MOSS string running along a channel line*

POINT	X	Y	Z	Chainage	Bearing			Radius of vertical curvature
3	6348.067	7679.230	175.854	870.000	115	54	48.5	660.000
4	6357.028	7674.792	176.210	880.000	116	46	53.7	660.000
5	6365.921	7670.219	176.562	890.000	117	38	58.9	660.000
6	6374.744	7665.511	176.910	900.000	118	31	4.2	660.000
7	6383.494	7660.670	177.254	910.000	119	23	9.4	660.000
8	6384.421	7660.147	177.290	911.064	119	28	42.0	660.000
9	6387.841	7658.201	177.424	915.000	119	48	18.2	723.254
10	6392.172	7655.701	177.593	920.000	120	10	37.4	823.527
11	6396.487	7653.175	177.762	925.000	120	30	2.9	956.079
12	6400.789	7650.627	177.929	930.000	120	46	34.8	1139.486

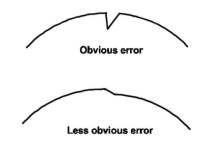

Obvious error

Less obvious error

POSSIBLE AMBIGUITIES IN INFORMATION TRANSFER

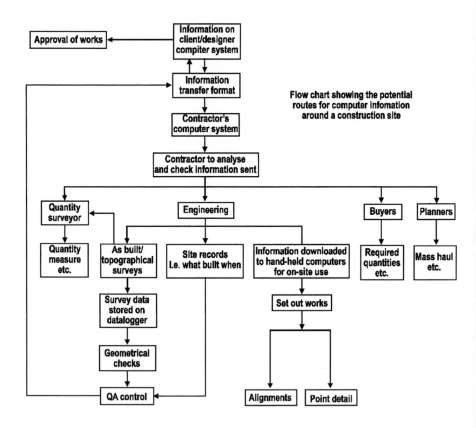

Flow chart showing the potential
routes for computer information
around a construction site

Appendix F: use of computers

The use of computers on site is now universal. The following text describes the principle uses and some of the implications for site engineers.

Design work

Most design work is carried out on a computer, the final details being transferred to paper for issuing to the contractor together with the relevant and CAD files. As site offices become less dependent on paper, documentation is often being issued as a file on disk, USB device, or transferred by email/electronic transfer. Among their many applications, computers are used for:

■ initial ground modelling
■ client's design
■ coordinated geometry and drawings production.

Issue of contract documentation

Contractor and designer should, ideally, use identical or interchangeable 'ground modelling and design' packages so that information issued is in a format immediately usable by both parties. Often this is not the case, and the designer must convert the information from the original format to one the contractor's computer system can understand.

Recovering the information

Because the information may be corrupted during conversion between formats, the engineer must scan through the received information for any obvious errors or ambiguities. Blatant errors can be dealt with immediately, but where the necessary correction is not obvious or is doubtful the engineer must confirm the requirements with the designer.

Uses on site

Once the information sent by the designer has been successfully downloaded into the contractor's system the potential uses are many. If controlled by a single competent person or department, the risk of information duplication and the introduction of errors and discrepancies is much reduced.

Engineering department

The engineering department is mainly interested in the information on the contract drawings. Whereas drawings can only show a two-dimensional representation, information contained in a computer-aided package can be interpreted three-dimensionally. The data will be produced as a series of alignments (eg for roads) or as point detail connected by straight lines (eg for structures) or a rotational 3D graphic models.

Data loggers and hand-held computers

Now commonly combined. Mainly used for setting-out and surveying purposes.

Data loggers

■ hand-held computers used on site for storing survey information, eg as-built surveys, or pre-calculated setting-out data
■ information saved can be downloaded into the computer system and comparisons made, eg between as-built and designed
■ survey information can be translated to the relevant format and issued back to the designer for 'Approval of the Works'.

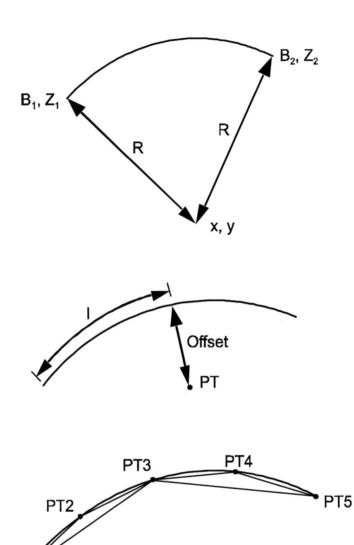

ALIGNMENTS

Hand-held computers

Given the coordinates and levels of primary setting-out stations hand-held computers with suitable programs can be used to calculate bearings and distances to generated or defined points, obviating the need for small calculators and field books, and with less likelihood of error.

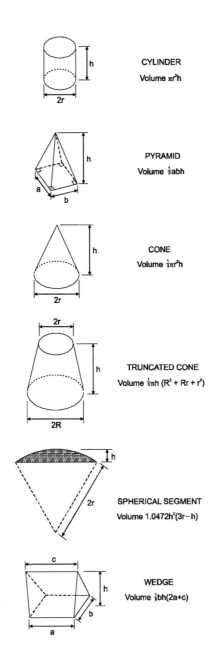

CYLINDER
Volume $\pi r^2 h$

PYRAMID
Volume $\frac{1}{3}abh$

CONE
Volume $\frac{1}{3}\pi r^2 h$

TRUNCATED CONE
Volume $\frac{1}{3}\pi h \, (R^2 + Rr + r^2)$

SPHERICAL SEGMENT
Volume $1.0472h^2(3r-h)$

WEDGE
Volume $\frac{1}{6}bh(2a+c)$

Appendix G: useful formulae

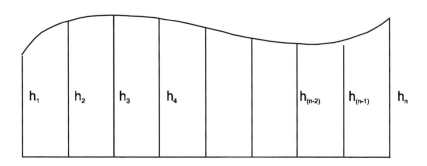

Simpsons: Area $= \dfrac{d}{3}(h_1 + 4h_2 = 2h_3 + 4h_4\ldots\ldots = 2h_{n-2} + 4h_{n-1} + h_n)$

where n is an odd number

Trapezoidal: Area $= \dfrac{d}{2}(h_1 + 2h_2 + 2h_3\ldots\ldots = 2h_{n-2} + h_n)$

Mean ordinate: Area $= \dfrac{\Sigma d}{n}(h_1 + h_2 + h_3\ldots\ldots h_{n-1} + h_n)$

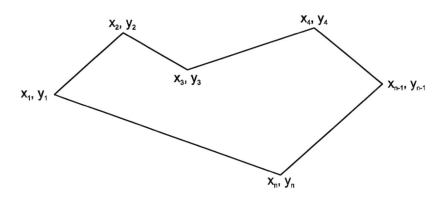

$$Area = \frac{1}{2}[(x_2 - x_1)(y_2 + y_1) + (x_3 - x_2)(y_3 + y_2) + \ldots(x_n - x_{n-1})(y_n + y_{n-1}) + (x_1 - x_n)(y_1 + y_n)]$$

Bibliography

General

ALLAN, A L (1993)
Practical surveying and computations, Second edition
Butterworth-Heinemann Ltd, Elsevier Science and Technology, UK. ISBN: 978-075060-393-5

BANNISTER, A and BAKER, R (1994)
Solving problems in surveying. Second edition
Longman – Pearson Education Ltd, Essex. ISBN: 978-058223-644-8

BANNISTER, A, RAYMOND, S and BAKER, R (1992)
Surveying, Sixth edition
Longman – Pearson Education Ltd, Essex. ISBN: 978-058207-688-4

SCHOFIELD, W and BREACH, M (1993)
Engineering surveying, Fourth edition
Butterworth-Heinemann Ltd, Elsevier Science and Technology, UK. ISBN: 978-075066-949-8

UREN, J and PRICE, W F (1997)
Calculations for engineering surveys
Van Nostrand Reinhold, New York. ISBN: 978-044230-583-3

UREN, J and PRICE, W F (2005)
Surveying for engineers, Fourth edition
Palgrave Macmillan, Basingstoke, Hants. ISBN: 978-140392-054-6

Specific topics

BUILDING RESEARCH ESTABLISHMENT (1980)
Accuracy in setting-out
Digest 234, BRE, Watford

BUILDING RESEARCH ESTABLISHMENT (1977)
Site use of the theodolite and surveyor's level
Digest 202, BRE, Watford

MEDICAL RESEARCH COUNCIL DECOMPRESSION SICKNESS PANEL (1982)
Medical code of practice for work in compressed air, third edition
CIRIA, Report 44, London. ISBN: 978-0-86017-175-1

POTTER, M (1995)
Planning to build? A practical introduction to the construction process
CIRIA, SP113, London. ISBN: 978-0-86017-433-2

BIELBY, S C (1997)
Site safety: a handbook for young construction professionals, second edition)
CIRIA, SP130, London. ISBN: 978-0-86017-455-4

COUNTY SURVEYORS' SOCIETY (1969)
Highway transition curve tables (metric)
Carriers Publishing Co, London. Available from Drydens (Printers) Ltd, 192 Brent Crescent, London NW10 7XU)

DEPARTMENT OF TRANSPORT (1991)
Manual of contract documents for highway works (MCHW), Volume 1 Specification for highway works (SHW)
HMSO, UK (with amendments August 1993, August 1994)

ICE/ICES (1997)
Management of setting-out in construction
Thomas Telford, London. ISBN: 978-072772-614-8

Good practice; accuracy and error corrections

IRVINE, D J and SMITH, R J H (1983)
Trenching practice
CIRIA, Report 97, London. ISBN: 978-086017-192-8

NATIONAL JOINT HEALTH AND SAFETY COMMITTEE FOR THE WATER SERVICE (1979)
Safe working in sewers and at sewage works
Health and Safety Guideline No. 2, Water Authorities Association, (withdrawn 1994)

BRITISH GAS PLC/UK WATER INDUSTRY ENGINEERING AND OPERATIONS COMMITTEE (1993)
Model consultative procedures for pipe construction involving deep excavation
NATIONAL WATER COUNCIL, ERS M60, UK

PRICE, W F and UREN, J (1989)
Laser surveying
Chapman & Hall, USA. ISBN: 978-074760-023-7

Standards

BS 4484: 1969 *Specification for measuring instruments for constructional works. Metric graduation and figuring of instruments for linear measurement*

BS 5606: 1990 *Guide to accuracy in building*

BS 5964: *Building setting-out and measurement* Part 1:1990 *Methods of measuring, planning and organisation, and acceptance criteria* Part 2: *Measuring stations and targets* Part 3: *Check-lists for the procurement of surveys and measurement services*

BS 7334: 1990, 1992 *Measuring instruments for building construction* Parts 1-8 BS EN 60825-1: 1994 *Safety of laser products. Equipment classification requirements and user's guide.*

Suggested colour code

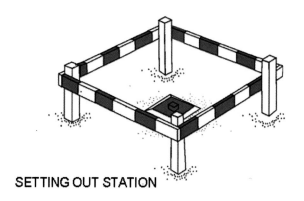

SETTING OUT STATION

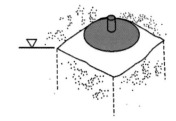

TEMPORARY BENCH MARK

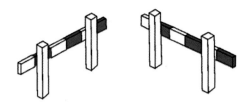

CORNER PROFILES

For grid lines

For offset lines

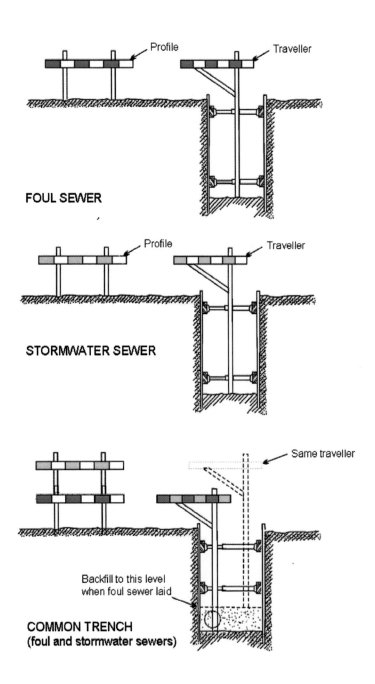

FOUL SEWER

STORMWATER SEWER

COMMON TRENCH
(foul and stormwater sewers)

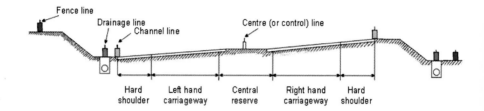

MOTORWAY OR DUAL CARRIAGEWAY (schematic)

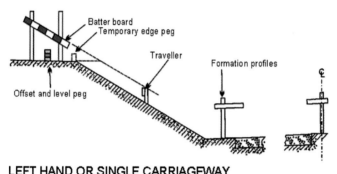

LEFT HAND OR SINGLE CARRIAGEWAY

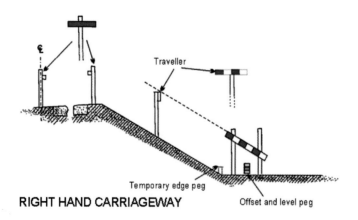

RIGHT HAND CARRIAGEWAY

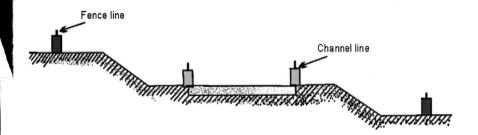

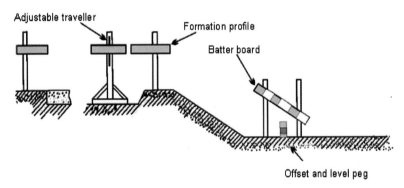

LINK AND SLIP ROADS

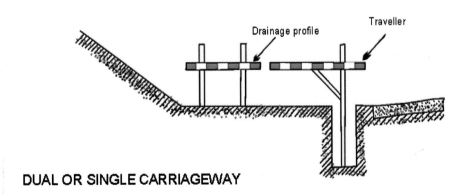

DUAL OR SINGLE CARRIAGEWAY